Two Californias

The Truth About
the Split-State Movement

TWO

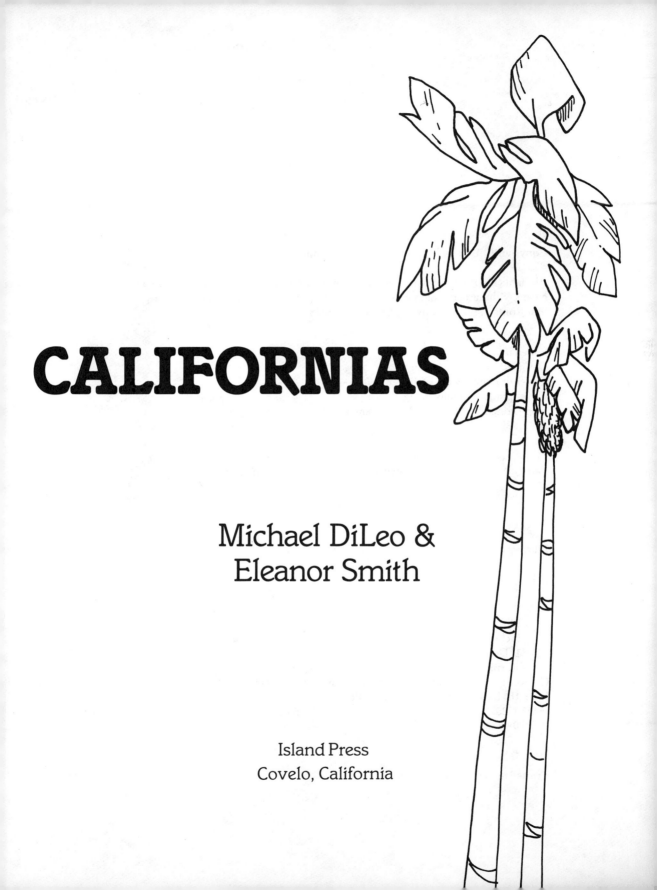

CALIFORNIAS

Michael DiLeo &
Eleanor Smith

Island Press
Covelo, California

Edited by Bruce Colman
Designed by Nancy Austin
Cover drawing by Bill Oetinger
Cover photo by Frank Wing
Copy edited by Mary Lou Van Deventer

Composed in ITC Italia and Paladium by Ann Flanagan Typography, Berkeley, and Patrick Miller, Berkeley. Printed and bound by Consolidated Printers, Berkeley.

Library of Congress Cataloging in Publication Data

DiLeo, Michael, 1946–
    Two Californias.

    Bibliography: p.
    Includes index.
    1. Decentralization in government—California.
2. California—Politics and government.   3. Regionalism—
California.   4. Water resources development—California.
I. Smith, Eleanor, 1954–   . II. Title.
JS451.C25D54   1983   320.9794   83-307
ISBN 0-933280-16-5 (pbk.)

Printed in the United States of America.

Contents

Foreword

When Jeremy Joan Hewes, a long time contributing editor to Island Press, suggested that we do a book on the concept of two Californias, the idea spun through our editorial offices provocatively.

At the time (spring '82), California was gearing up for a vote on building the Peripheral Canal, and splitting the state in two was a possibility that appeared with increasing frequency (and seriousness) in the media. It popped up even more often around Covelo, the small, Northern California town that is the site of the Island Press editorial offices. If California voters approved construction of the canal, which seemed likely at the time, the Press offices stood an excellent chance of joining the rest of Round Valley at the bottom of a lake. California water interests' next move would almost inevitably be to build a dam on the Eel River at Dos Rios, in order to keep the canal full.

Split the state? Not a bad idea, from the standpoint of a frustrated and seemingly powerless Northerner.

We were fortunate that Eleanor Smith and Michael DiLeo came forward to write *Two Californias*. Both are talented researchers and agreed to keep an open mind about splitting the state while they wrote the book.

Eleanor, who was born in Orange County, spent most of her childhood on rural Long Island. She moved back to Newport Beach for junior high school, high school, and college. After graduating from the University of California at Irvine, in 1976, she moved to the Bay Area and worked as managing editor of *Not Man Apart*, published by Friends of the Earth. She left FOE in 1981, and has since done investigative reporting for a variety of regional and national magazines. She now lives in Berkeley.

A native New Yorker, Michael graduated from Duke University in North Carolina in 1968 and migrated to UCLA, where he received his MA in history. After that, he moved north to the Bay Area and on, for a time, to the rugged North Coast mountains, where he was active in environmental work. He currently lives in Marin County and writes history and government text books for Harcourt, Brace, Jovanovich.

Although an affirmative answer to the split-state question had intriguing appeal, we felt it was premature. Discussions that went beyond the obvious one-liners and utopian scenarios had quickly uncovered a startling complexity of serious issues. We didn't expect Michael and Eleanor to be unbiased (who, having lived in this complicated state longer than three weeks, could genuinely have no opinion on the questions involved?), but we did hope that their biases would work creatively.

As you will see, that is just what happened. The authors have brought the extraordinary range of their own experiences to bear. Both have lived long periods in both parts of California (or, if you will, in both Californias); both have hands-on experience with some of the difficult environmental issues that face the state; and both have, in addition, points of view influenced by time and residence in other parts of the country.

These qualities became increasingly important as the book took shape. An idea that, on its surface, captures the imagination, became a multi-faceted metaphor with echoes back into California's history and forward into our collective future. The urge to split the state raises some very basic questions about the ways we choose to live together. Or about the ways we choose to live apart. These questions always remain, in the most fundamental sense, open.

Barbara Dean
Executive Director, Island Press
Covelo
February 1983

Preface

Two Californias: the idea that a deep rift exists between Northern and Southern California permeates the folklore of this sprawling state. Millions of people believe wholeheartedly that there *are* two Californias (they are talking about culture). Millions of other people are convinced that there *should be* two Californias (in political terms). Of course, there are countless others who think nothing of the kind. California as it is and as it should be is something that, like beauty, is in the eye of the beholder.

Both of us began this book with a strong sense that splitting the state might actually solve California's problems. In our early research we uncovered a spate of similar secession movements around the land. Canada, of course, is besieged by separatist forces in Quebec, where a question of cultural integrity is involved, and to a lesser extent in the western provinces, where money and natural resources lie at the heart of the problem.

In the United States there may be a dozen such efforts. Separatism seems to have become a full-blown craze. In the Florida keys, fishermen have proposed forming a new state called the Conch Republic. Their biggest gripe appears to be a roadblock placed on the highway to slow the influx of illegal aliens. This, say the fishermen, slows only the flow of traffic and the hauling of seafood to market. In Alaska, some real estate developers have proposed secession to escape annoying restrictions placed by environmentalists from the Lower 48. Chicano radicals in the Southwest want to form the state of Aztlan from the area between San Diego and the Platte River. In Texas a movement led by former state senator Bob Gammage and state representative Dan Kubiak seeks to create five modestly sized states out of one big one. The idea here is to quintuple Texas's relatively meager representation in the US Senate. (This movement may technically be the most feasible of all; see chapter 5.) Among the other proposed new political entities are the Republic of South Jersey, the Republic of Nantucket, the Republic of Martha's Vineyard, and the Superior Republic, in the far upper Midwest.

What's going on here? we asked ourselves. Is secession a cop-out, a fad, a cheap publicity gag, an American atavism left over from the Civil War, or is it an important political current? Though unqualified to speak on situations in other states, we have concluded that the North-South split in California is serious business. We have three reasons for this opinion.

First, there is the history factor. As you will discover in the reading of this book, the idea of splitting California into two or more states has a long and colorful history, more than a century and a half in duration, beginning before California was even a state.

Secondly, there is the magnitude of the issues involved. The split-state idea has always served as a vehicle, a means of expressing sentiment about some other issue. In the late nineteenth century, the problem was economic disparity; in the 1940s bad roads caused a rebellion; in the 1960s, legislative apportionment was at issue. Today, the issues are as fundamental as can be: water, quality of life, and humankind's relationship to the natural world. In contemporary California, secession is less an attempt to draw the attention of an indifferent public than it is a desperate recommendation for solving a nearly unsolvable set of problems.

Finally, there is the unique biogeographical structure of the state itself. Virtually every state in the Union has some mild form of internal division. This rift is typically between big city and country, between upstate and downstate, between capital (or trade center) and cow county. As the political nexus of power has shifted from farm to city over the last century, the disenfranchised rural inhabitants have had to scratch and claw and occasionally raise the cry of secession to retain any voice in state politics.

In California, the situation is quite different. The two feuding regions are loci of considerable power. The North has valuable farmland, timber, water, a growing high technology industry, one of the world's leading ports, the state's banking and commercial center, respected institutions of higher learning, a thriving tourist industry, the state capital, and one of America's largest metropolitan areas surrounding San Francisco Bay. That is not, as they say, chopped liver. The Southland has America's second largest city, an economy bigger than many countries', agriculture, high tech industries, the movie business, and most important of all, well more than half the state's population in a sprawling megalopolis which extends from Los Angeles nearly to the Mexican border. This is not a bunch of outmanned rubes fighting a stacked deck of power brokers. These are two worthy adversaries, each of whom means business.

While the rivalry between North and South California has serious implications, it does also have its lighter side. We want to capture that aspect, as well. In this book, we are after, if you will, the right-brain and the left-brain of California. We want to include the facts and figures, the policy statements and the brutish historical data; but equally important to our purpose is the subtext, the shadow-world of cultural myths and interregional rivalry. The snide humor of the North-South rivalry is as significant in its way as the loftiest debate on water policy.

The more we learned about California's peculiar bifurcation, the farther we moved from our original conception of the book. We have not written a political tract recommending a split-state, although a lot of how-to information is included for those so inclined. We have not exactly written a history of California, either, although tracing the North-South split through time takes one into the nooks and crannies of the past. We began this book primarily as an informative piece for Californians; what we have composed, it seems, is the complete anatomy of a feud, a venerable, deep-running, free-swinging, socio-political mother of a feud. We think our research has uncovered most of what divides North Californians from South Californians; but at the same time, we hope our reflections aim toward a vision of what may keep us all together as human beings. Follow along, then, as we begin at the beginning in our tale of two Californias.

We would like to express our deepest appreciation to the following people for their assistance and cooperation: Robert Arnold, of the Center for the Continuing Study of California's Economy; Ann Bailey, Greg diGiere and the staff of Senator Barry Keene's office; Kathleen Bartoloni, Peter Berg, Ernest Callenbach, Frank Gleason, Tom Graff, Jeffrey Knight, Clifford Lee, David Nesmith, Steve Pitcher, Carl Pope, and Marc Reisner; Jack Smith of the California Farm Bureau Federation; Don Villarejo of the California Institute for Rural Studies; and Betty Doyal for sending the *Fresno Bee* clips. Special thanks go to Harry Dennis, Richard Clark, Ilona Inovsky of the Survey Research Center, and the staff of the University of California at Berkeley's Water Resources Center Archives, for their most gracious research assistance; to Bruce Colman, our editor; to Barbara Dean for her patience and support; to Frank Wing for his generosity; to Stephanie Mills for her most appreciated contribution; and to Jeremy Hewes for the initial inspiration.

Eleanor Smith &
Michael DiLeo

Chronology of
Split-State Movements

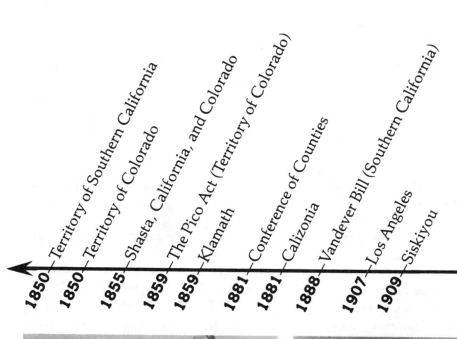

1850	1850	1855	1859	1859	1881	1881	1888	1907	1909
Territory of Southern California	Territory of Colorado	Shasta, California, and Colorado	The Pico Act (Territory of Colorado)	Klamath	Conference of Counties	Calizonia	Vandever Bill (Southern California)	Los Angeles	Siskiyou

ER AT PALA MISSION AND MEXICAN VAQUEROS

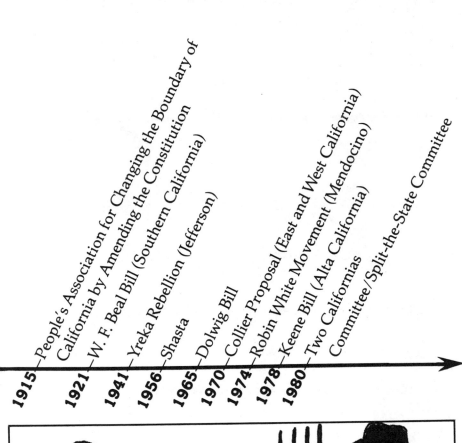

1915 — People's Association for Changing the Boundary of California by Amending the Constitution
1921 — W. F. Beal Bill (Southern California)
1941 — Yreka Rebellion (Jefferson)
1956 — Shasta
1965 — Dolwig Bill
1970 — Collier Proposal (East and West California)
1974 — Robin White Movement (Mendocino)
1978 — Keene Bill (Alta California)
1980 — Two Californias
        Committee/Split-the-State Committee

7

1
The Golden State 1820–1914

ishing he'd never come to Monterey, the new governor of Alta California looked out on a dark, surly ocean. A blast of wind chilled his thin, juiceless body. He sighed and started to cough, cursing the fates that had brought him here. The year was 1825.

Jose Maria Echeandia was brooding about more than his health and the weather. The love of his life, the dazzling Senorita Josefa Carrillo, was living in San Diego, hundreds of miles to the south. While the governor attended to the tedious affairs of state, he wondered if some young suitor was bidding for her heart. Just the thought made him begin coughing again.

He gazed out once more onto the stormy Pacific. The weather seemed so oppressive. The swirling fog was turning the misshapen conifers along the coast into strange apparitions. He hated this gloomy place.

On that wintry day in 1825, Echeandia, the third Mexican governor of Alta California (the territory north of modern-day Baja California), made a decision: he would leave Monterey and move to San Diego. There was just one problem. Monterey had been the seat of the territorial government since 1775. That didn't stop Echeandia, though; he simply took the government with him. Upon his arrival in San Diego, a short time later, he named that town the new capital.

The residents in and around Monterey, called *arribenos* ("uplanders"), were furious. They felt betrayed and abandoned. The *abajenos*, residents of the southern part of the territory,

> **"A division of the state into two or more states is a political necessity which will be recognized by all parties sooner or later."**
> **—Sacramento Union, 1853**

were delighted. By this move south, Echeandia planted seeds of bitterness, seeds that would produce armed conflict a few years later.

Thus was born one of the first of many feuds between the northern and southern halves of the place we call California. For more than a century and a half, this rivalry has continued. Its ebbs and flows have played a large part in shaping the destiny of what is now America's largest state.

Two Californios

Actually, the idea of dividing California in two arose as early as the eighteenth century. The Franciscan fathers proposed that the Spanish territory of Alta California be severed at the Tehachapi Mountains, at a point just north of Santa Barbara. The padres felt that the region, which extended some 600 miles from San Diego to Sonoma, was simply too large to administer as one unit. Because the area was so sparsely populated, the plan was rejected as unnecessary.

Although he did not win the heart of Josefa Carrillo (she ran off with an American sea captain), Governor Echeandia won the praise of many "Californios"—as the Mexicans who lived in the remote district were known—for his efforts to secularize the wealthy, omnipotent missions. His mistake was to foolishly let his soldiers—former convicts sent to this distant backwater as punishment and who had stayed in Monterey—nearly starve to death. In 1828, the army rebelled and headed south to oust him. The insurgents met the governor's men at Santa Barbara and were forced to withdraw. Echeandia soon recaptured Monterey and sent the rebellious soldiers back to Mexico. Then the governor retired.

Rebellions, uprisings, conspiracies, and intrigues colored the entire Mexican era in California, from the time of liberation from Spain in 1821 to the American conquest in 1847. The Californios resented the arrogant, self-serving governors appointed by Mexico City. Because the federal capital was so far away, they felt little fear of reprisal if they took up arms against these incompetent, often ruthless rulers. Eventually, some group or other would brandish their weapons against each new appointee and send him scurrying back to Mexico. After Echeandia, California had nine governors in 15 years.

Californios also fought with each other. "Los Angeles, which by then was the most populous pueblo in the territory, kept trying to snatch from Monterey the capital and the customs house with its essential revenues," writes David Lavender in *California: A Bicentennial History*.[1]

In 1831, Echeandia overthrew his successor, the militant Manuel Victoria.

He installed himself as governor, for the second time, in San Diego. Stewing since his first administration, the *arribenos* of Monterey induced Agustin V. Zamarano to challenge the impudent *abajeno* and declare Monterey the capital. Each leader marshalled his troops for battle but eventually compromised by dividing California into two territories.

Echeandia ruled the southern half, which extended as far north as the San Gabriel Mission. Zamarano wielded the scepter of power in the north, the southern limit of which was the San Fernando Mission. In between the two territories lay a no-man's land. If one side's army crossed over the line, the other would instantly begin a war game.

After a year of this, Mexico City sent Jose Figueroa to govern California. Figueroa united the torn territory and made Monterey his official residence. "For a time peace reigned, but the end of the controversy was not yet—the politicians of the south were placid, but they were plotting," historian J. M. Guinn wrote.[2]

In fact, one prominent Angeleno persuaded the Mexican Congress to declare Los Angeles the capital of both Baja and Alta California. The citizens of the southern pueblo, overjoyed, sent a request to Monterey for the government archives and for the governor to move at once to the new capital.

Taking umbrage, the northern politicians responded with a demand of their own. They insisted that a suitable *palacio* for the governor be found before the government moved to Los Angeles. Representatives were dispatched to find one, but to no avail. The officials stayed in Monterey and proceeded to taunt Angelenos with "invidious comparisons"—their lack of polish, their provincialism, and their poverty.

In 1844, Pio Pico and his followers ousted the last Mexico-City-appointed governor of California, the arrogant Manuel Micheltorena. A native of Southern California, Pico declared himself governor and Los Angeles the capital. But Monterey kept the coveted customs house; there Jose Castro, the new military commandant, ruled as if ultimate power were his. Within two years it was clear the two Californios would cross swords.

In 1846 Pico set out with a small contigent of *abajenos* to teach Castro and the plotting politicians of Monterey a lesson. The group got as far north as San Luis Obispo when it encountered a messenger. With shock, Pico learned that an American, Commodore Robert Sloat, had captured Monterey and taken possession of California in the name of the United States. He retreated to Los Angeles. The war between the *arribenos* and *abajenos* was over; the two factions united in the face of a common enemy, the Yankees.

Halcyon Days

From 1846–1847, the Californios fought valiantly against the invading Americans. In one particularly bloody battle, near the Indian village of San Pasqual (near San Diego), General Stephen Kearny's force encountered the troops of General Andres Pico, brother of Governor Pico. The Americans had marched all the way from Fort Leavenworth, Kansas, and were exhausted. In spite of this, a cold mist, and a fog-shrouded and unfamiliar battlefield, Kearny attacked Pico's men early on the morning of December 6, 1846.

"The Americans were no match for the lance-equipped and superbly mounted Californios," wrote historians Beck and Williams.[3] Kearny lost 21 men with 18 wounded, but Pico lost only one man. This was the most significant battle ever fought on California soil—and also the last victory for the Californios.

A month and a half later, Andres Pico—whom we'll meet again later—surrendered on behalf of all his compatriots to General John C. Fremont. The Cahuenga Capitulation ended the hostilities in January 1847. The following year, the United States got what it had been after for several years: the Treaty of Guadalupe-Hidalgo formally ceded California to the US. Between 1848 and

Pala Mission. Courtesy, the Bancroft Library

236. THE BELL TOWER AT PALA MISSION AND MEXICAN VAQUEROS.

1850, the interim preceding statehood, a series of military governors, appointed by Washington, ruled the far western territory. Much like their Mexican predecessors, they argued constantly over who ruled which part of the state.

When the Americans took over in California, they discovered a life-style wholly different from their own. While a few Californios had taken arms against their oppressive governors, the incessant rebellions did little to interfere with the lives of the majority, who lived on the ranchos—large cities unto themselves. This was, according to historian Robert Glass Cleland, "the day of the Dons, an idyllic interlude during which a people of simple wants, untroubled either by poverty or by the ambition for great wealth, gave themselves over wholeheartedly and successfully to the grand and primary business of the enjoyment of life."[4]

The ranchos were actually former mission holdings, which had been divided up as part of the secularization plan. The Mexican governors granted large chunks of land to a number of Californios and to a few foreigners. To qualify for a grant, the applicants had only to promise to use and stock the land with at least 2000 head of cattle. Between 1834 and 1846, more than 800 land grants containing 13 to 14 million acres—about one-fourth of the land area in California—were authorized. Most were in Southern California.

Rancho life was easy-going. Cattle grazed on open ranges, so there was little fence building and repair to be done; mission-raised Indian laborers did the mundane work; and a few *vaqueros* handled the branding and butchering.

Raising cattle for hides and tallow was extremely profitable. "Each year, the Californios sold, as a group, some 75,000 hides for an average of $2.00 each," states Lavender.[5] Very little of this income was taxed for the support of education, social services, transportation, or defense. As a result, the rancheros and their families lived the life of Riley. Their huge profits went for "fancy suits emblazoned with gold and silver braid, showy cowhide boots made in Boston, ornate saddles, and broad-brimmed beaver hats. For the women, there were satin slippers, rustling petticoats, and *rebozos* of Chinese silk that might cost $150 or more apiece."

All That Glitters

One of the foreign ranchers enjoying these halcyon days was John Sutter, a Swiss immigrant whose vast settlement in the Sierra foothills near Coloma had become a refuge for early travellers crossing the mountains into California. Sutter was called "the baron of Sacramento Valley," and with his associates, raised stock, farmed, traded, and harvested timber at Sutter's Fort. One fateful decision Sutter made was to build a sawmill on the American River to process the fort's timber.

James W. Marshall, a skilled wheelwright and itinerant carpenter, had been named foreman of the sawmill crew. To broaden the channel where the water would supply the mill with power, the crew allowed the river to run through a tailrace at night. One morning in January 1848, Marshall recalled later, "I went down as usual and after shutting off the water from the race, I stepped into it near the lower end, and there upon the rock, about six inches beneath the surface of the water, I discovered the gold."[6]

Marshall tested the golden flakes to make sure they were not fool's gold. Satisfied, he told Sutter about his finding. Sutter swore his foreman and everyone else at the fort to secrecy. Then he sent two employees to Monterey to ask the military governor to grant him the piece of land upon which the sawmill was being built. The governor refused and sent Sutter's envoys back to Coloma. But the secret was out.

Sluice boxes used to trap placer gold during the 1850's

Within weeks, prospectors from all over Northern California began to arrive at Sutter's Fort. Trespassers butchered Sutter's cattle, overran his lands, and stole his property. Not nearly as lucky as others, who carted away barrels of gold, Sutter and Marshall fared only moderately well as miners. Sutter ultimately went bankrupt, the victim of several unwise business deals.

The discovery of gold in California rocked the world. Like wildfire, the news spread across the country and leapt over the oceans. From as far away as France and China, not to mention the East Coast and the Midwest, men dropped their guns, their pens, their plows, or whatever else they were holding at the time, and rushed off to the goldfields of the Sierra Nevada. They kissed their wives goodbye and promised to be home soon with a sackful of nuggets that would end all their worries forever. Single men left in a jiffy, with promises to no one but themselves to strike it rich fast. They traveled across the Great Plains in wagon trains, or took boats to San Francisco, then hurried to the foothills. Of course, those who already lived in California had the advantage of getting to the goldfields first to stake out claims. In 1848, nearly all of the residents of San Francisco and Monterey up and left; armed with shovel and pick, they headed east to the Mother Lode.

Shortly after Sutter's men visited him in Monterey, Colonel R. B. Mason, military governor of the conquered province of Alta California, decided to see for himself what all the commotion was about. In the summer of 1848 he visited a site along the American River called the Mormon Diggins. He described the scene in a letter to the Secretary of War in Washington:

"The hillsides were thickly strewn with canvas tents and bush arbours; a store was erected, and several boarding shanties in operation. The day was intensely hot, yet about two hundred men were at work in the full glare of the sun, washing for gold—some with tin pans, some with close woven Indian baskets, but the greater part had a rude machine known as the cradle....Four men are required to work this machine; one digs the ground in the bank close by the stream; another carries it to the cradle and empties it on the grate; a third gives a violent rocking motion to the machine; whilst a fourth dashes on water from the stream itself....A party of four men thus employed, at the lower mines, averaged $100 a day."[7]

The situation out west dismayed President Polk. He informed Congress in late 1848: "The effects produced by the discovery of these rich mineral deposits, and the success which has attended the labours of those who have resorted to them, have produced a surprising change in the state of affairs in California. Labour commands a most exorbitant price, and all other pursuits but that of

searching for the precious metals are abandoned....Ships arriving on the
coast are deserted by their crews, and their voyages suspended for want of
sailors. Our commanding officer there entertains apprehensions that soldiers
cannot be kept in the public service without a large increase of pay. Desertions
in his command have become frequent."[8]

San Francisco soon became a bustling city where miners bought their supplies and returned with their gold. They exchanged it for needed goods or
squandered it in the gambling halls or on the fancy ladies who arrived soon
after the deluge of unattached men. The city's population swelled to nearly
20,000 by the end of 1849; most were men between the ages of 18 and 40.

One of these men was Bayard Taylor, a bright, young reporter for the *New
York Tribune,* sent west on assignment. Taylor reached San Francisco in
August of 1849. He described the scene as he disembarked:

"A furious wind was blowing down through a gap in the hills, filling the
streets with clouds of dust. On every side stood buildings of all kinds, begun or
half-finished, and the greater part of them were canvas sheds, open in front,
and covered with all kinds of signs, in all languages....The streets were full of
people, hurrying to and fro, and of as diverse and bizarre a character as the
houses: Yankees of every possible variety, native Californians in sarapes and
sombreros, Chilians, Sonorians, Kanakas from Hawaii, Chinese with long
tails, Malays armed with their everlasting creeses [swords], and others in
whose embrowned and bearded visages it was impossible to recognize any
especial nationality. We came at last into the plaza, now dignified by the name
of Portsmouth Square."[9]

San Francisco had telescoped a half century's growth into one year, Taylor
would remark later. Within five years of the discovery of gold, the City by the
Bay rivaled New York financially and Boston culturally. The city's per capita
wealth was the highest in the nation, although not at all evenly distributed.
Hotels charged outrageous sums for rooms; water supplies, never before taxed
by the city's population, had to be hauled in by barge from Marin County to
meet San Francisco's urgent needs.

Sunset on the Day of the Dons

The discovery of gold effectively divided California into two states. While
the din of pick striking rock filled the valleys of the Sierra foothills and San
Francisco grew into a full-fledged city, Southern California slumbered on.

Prior to 1848, most of the people who lived in California had lived in the
South, primarily on the ranchos. The discovery of gold shifted the centers of

Los Angeles in 1857. Courtesy, the Bancroft Library

population to the North. At the beginning of 1849, some 26,000 people called California home—about half of them Californios, the rest newcomers. By the end of that year, the state's population had jumped to 128,000—the majority miners in the North.

Benjamin Butler Harris, a prospector from Texas, rode overland by horseback in 1849. Crossing the desert, he and his companions entered the territory from the south and paused briefly in Los Angeles on their way to the goldfields. Harris described the town as containing "2000 to 3000 people....Its houses were one-story adobe—roofed with thatch smeared over with brea which during the heat of day dropped from the eaves like ropy tar. Only three or four Americans lived there—one was Don Benito Wilson at whose store we replenished supplies for the journey to the mines."[10]

Initially the rancheros prospered from the influx of miners needing food and supplies, which they provided. Within a few years, though, the bonanza in the North was over. The 49ers had gleaned nearly all the gold that centuries had scattered along riverbanks and lodged into bedrock. Disappointed and either too poor or too ashamed to go home, many failed miners headed south to try their luck at farming. Several took to squatting on rancho land, and, being Yankees, pressed the US government to give them some of the lands held by the Californios.

If American settlement had proceeded as slowly as it began, historian Lavender asserts, the two cultures might have assimilated peacefully. But the Gold Rush prevented this. "California was inundated by exploiters who accepted Spanish place names because they were convenient, borrowed Mexican mining customs and techniques because they were useful, and studied Mexican law because it helped them challenge the validity of the land grants. Otherwise the conquerors brushed aside the Hispanic ways and people as irrelevant."[11]

The Yankees Call for Order

With the massive infusion of American and other fortune-hunters into California, it was not long before many felt the need to set up some form of civil government. Lawlessness was rampant in the mines as well as in the cities. Vigilante gangs became the judge and jury, dispensing justice quickly and often mercilessly. The military governors bickered over each other's jurisdiction, while the Americans grew impatient with the Californio legal system.

Some of the politically minded newcomers wanted to see California admitted to the Union so they could seek office in the new state government, but Congress was dragging its feet. Slavery had already incited passionate debate in Washington and it was feared that adding another state at this time would upset the delicate balance of free and slave states in Congress. There were 15 of each at that point.

Rather than wait for Congress to settle its differences, Mason's successor as military governor, General Bennet Riley, called a convention to meet in Monterey for the purpose of forming a constitution for California.

Representatives from all over the territory gathered in a stone school building in the seaside capital during September of 1849. In a room 30 feet by 60 feet, the 48 delegates set to work framing the government. The group was relatively young, though rich in experience. A half-dozen were Californios and twelve were from the South. The delegates hailed from a wide range of occupations: twelve were lawyers, ten were farmers or rancheros, seven were merchants, and several were military officers. Thomas O. Larkin, former US consul in Mexican Monterey, was there, as were Henry W. Halleck, who later became a prominent member of a prestigious law firm, and Elisha Oscar Crosby, a transplanted New Yorker who had with him a copy of New York State's constitution to use as a guide. Also present was Dr. William M. Gwin, a former US Congressman from Mississippi and friend of Andrew Jackson. He had come to California with the express purpose of becoming the first US senator from the new state. A dapper man, Gwin preened for the chairman-

William Gwin, one of California's first Senators.

ship of the convention but lost out to Dr. Robert Semple, a dentist, printer, and Bear Flagger.

It quickly became obvious that the Californios were not at all pleased with the prospect of tying their destiny to that of the uncouth, aggressive Yankees in Northern California. Delegate Gwin wrote about the convention several years later: "When they met to organize, the members showed a strange distrust of the motives of each other from various sections."[12]

The feelings of the Californios were not really so strange. They knew the American propensity for taxing real estate. They were certain the newcomers —now the majority of the population—would place an unfair tax burden on their extensive land holdings to support the state government. Since the Yankee newcomers leased their land from the federal government, rather than owning it, the Californios' fears were not unfounded.

After discerning that most of the delegates favored statehood, the Californios made a grandstand play. Sr. Jose Antonio Carrillo, a representative from

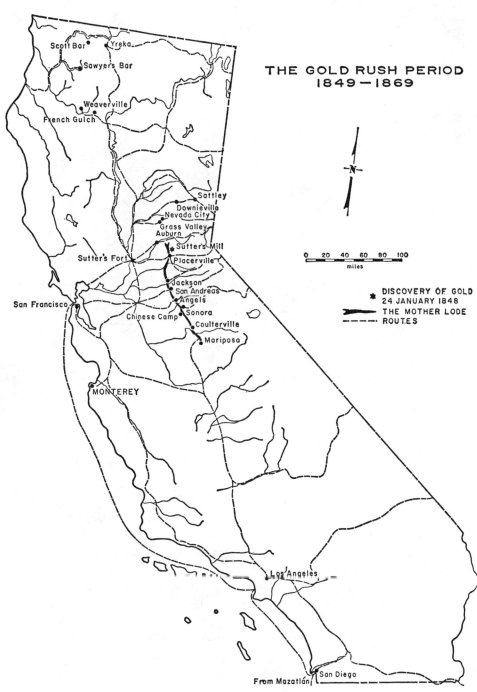

THE GOLD RUSH PERIOD
1849—1869

Scott Bar
Yreka
Sawyers Bar
Weaverville
French Gulch

Sattley
Downieville
Nevada City
Grass Valley
Auburn
Sutter's Mill
Sutter's Fort
Placerville
Jackson
San Andreas
Angels
Sonora
Chinese Camp
Coulterville
Mariposa

San Francisco

MONTEREY

Los Angeles

From Mazatlán   San Diego

0   20 40 60 80 100
miles

★ DISCOVERY OF GOLD
    24 JANUARY 1848
    THE MOTHER LODE
- - - ROUTES

From California, A History of the Golden State

Santa Barbara, proposed that the territory be divided in two by a line running east from San Luis Obispo. That way the area north of the line could have a state government and the area to the south could be administered under territorial status, as the Californios desired. The delegates, primarily Yankees who lived in the North, rejected his plan.

The Californios were livid. They threatened to quit the convention and leave the thorny problems of drafting a constitution to the newcomers. To prevent such a move, the delegates agreed to a compromise. They adopted a provision placing control of property tax rates in the hands of county tax assessors and boards of supervisors elected by the landowners themselves. This was designed to ensure that the rancheros would not be taxed oppressively.

While they were in an agreeable mood, the delegates approved a plan to divide the territory into 27 counties, based roughly on population densities. Because the Southland had so few inhabitants except along the coast, and the North "teemed with the transient population of the gold placers," as historian Rush put it, the territory east and northeast of San Francisco Bay was cut up into many small counties. Thousands of square miles to the south were lumped into two enormous counties, Mariposa and San Diego. Four smaller counties were created along the coast from Monterey to the Baja California border.[13]

To the surprise of many delegates, the hottest topic of debate was where the eastern border of the new state should be. Everyone agreed that the northern and southern borders should be where they are today, but the question of how far east the other border should be raised havoc. Some delegates suggested the new state encompass much of present-day Nevada and Utah. The brainchild of the ambitious William Gwin, "a tall and distinguished man who looked as if he should wear a toga,"[14] the plan was to get the delegates to make the Rocky Mountains the eastern boundary, not the Sierra Nevada and the Colorado River, as most of the delegates preferred. If he could accomplish that, Gwin would then have only to persuade Congress to divide this huge territory along the 36°30′ parallel, the "color line," to have two states—one free and one slave. A dyed-in-the-wool slavery advocate, Gwin was loyal to his native Mississippi's institutions and customs throughout his long career on the western frontier.

Promising the Californio delegates land grants larger than those they'd received from the Mexican governors, Gwin convinced them to go along with the scheme. It also happened to jibe with their desires to be separate from the golddiggers in the North. After a third reading, the delegates approved Gwin's proposal. All that was needed now was to authorize it as part of the constitution.

The northern delegates, mostly Yankees from free states, were in shock. They made a last-ditch effort to block Gwin's plan. While one group distracted the delegates in the hall, another ran off to find Francis Lippitt, an influential delegate from San Francisco who had missed the first vote because he suffered from malaria. Upon hearing of Gwin's plan to make Southern California a slave state, Lippitt swallowed a formidable dose of laudanum—tincture of opium—and set off for the convention hall.

He gave a stirring speech against the Gwin plan, but Lippitt later told *Century* magazine: "As to what I said I have not the slightest recollection, except that I dwelt earnestly on the improbability of the admission of a new free state covering such an immense territory as the Rocky Mountains boundary would give us, in view of the fierce and persistent opposition it would encounter from the Southern members in Congress. I was afterwards warmly complimented for my speech; but I have never taken any credit to myself for it, well knowing that whatever there may have been effective in it was due to the influence of the narcotic I had taken."[15]

The delegates voted down Gwin's plan and finally agreed that the new state should outlaw slavery. Shortly thereafter, they completed their work and submitted the new constitution to Congress, requesting admission to the Union as a single state. Congress was still wrestling with the slavery question and could not agree on what to do about California. So it stalled.

The Californios tried again to divide the territory, and Southern members of Congress even proposed a plan similar to Gwin's, but all suggestions to split California prior to admission fell by the wayside. Finally on September 9, 1850, Congress admitted California, with its current boundaries and free-state constitution, as the thirty-first state in the Union.

Californios Want Out

Back in the new state, miners in the North were busy scouring Sierra streams for bits of the glittering yellow metal. At the same time, American settlers—newcomers hoping to make a new life for themselves on the bountiful frontier as well as failed golddiggers—were squatting on rancho land. When the rancheros demanded they leave, the squatters went to their friend, newly elected US Senator William Gwin, for help. Although the land titles to the ranchos were to be "inviolably respected"—according to terms of the Treaty of Guadalupe-Hidalgo—the Americans successfully lobbied Congress to pass the Land Act of 1851. The law required grantees to prove that their claims were valid and that they covered the territory the Californios said they did. More than any

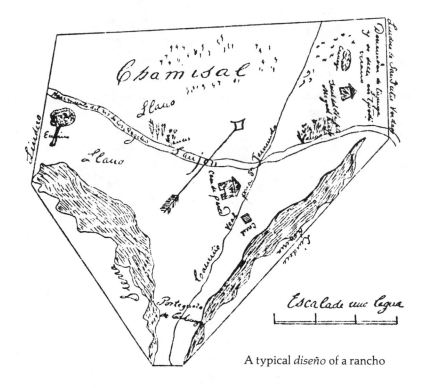

A typical *diseño* of a rancho

other assault on their way of life, the Land Act dealt a body blow to the Californios.

Proving a claim, it turned out, was no simple task. American lawyers took advantage of the different styles of the US and Mexican legal systems. The Mexican government had been less exacting in its requirements than the US was. One historian notes that, besides the cattle rule, the Mexican governors had required only that a petitioner for a grant describe it in his petition; submit a personally drawn *diseno,* or map; and demonstrate his personal commitment to ownership by walking about on the piece of land in front of witnesses, "uprooting grass, scattering handfuls of earth, tearing twigs, and performing other acts and ceremonies of real possession." Finally, he was to build a house on the land within a year and occupy it.[16]

Such vague language allowed for ill-defined boundaries, a number of discrepancies, and overlapping claims. The Americans also found the Mexicans' emphasis on ritual ludicrous and vulnerable to legal scrutiny.

The Land Act accomplished what its backers had intended it to. Within a few years, many rancheros were surrendering the bulk of their land to Yankee lawyers in exchange for vain attempts to defend their claims. Others were forced to sell their historic ranchos at bargain basement prices to pay exorbitant legal fees. "One out of ten of the bonafide landowners of Los Angeles County was reduced to bankruptcy by the federal land policy," Carey McWilliams points out. "At least 40 percent of the land owned under Mexican grants was sold, by the owners, to meet the costs and expenses involved in complying with the Act of 1851."[17]

Many rancheros became day laborers in the towns. Those who managed to hang onto some of their land did their best to keep up with the international beef trade. What the Land Act did not accomplish in 1850s, nature did in the following decade. Thousands of cattle drowned during a monstrous flood in 1862, and thousands more died during a severe drought in 1863-1864. Many of the remaining rancheros were forced to sell their land, for as little as ten cents an acre.

In addition to the Land Act's next-to-impossible requirements, the landholding Californios suffered other injustices at the hands of their new leaders. Just as they had predicted at the first constitutional convention, the expenses of the new government were weighing heavily on them. Because they owned most of the private property in the state, at least for a few years after admission, the Californios paid the bulk of taxes. Meanwhile, the miners in the North, who paid little or no property taxes, were free to dig their claims and enjoy their fortunes.

The Californios were not used to paying property taxes; the Mexican governors had taxed only brandy and wine to support their administrations. And the rancheros had been accustomed to bartering for goods and services, not paying cash. Thus, unfamiliar with both the concept and practice of paying property taxes and with dealing in a monetary economy, many rancheros felt powerless as the rising American tide swept away their sense of self-determination.

In August 1851, the Californios were again organizing a drive to split the state and declare the Southland a territory independent from the northern state. They were sick and tired of the Yankees' abuses. They wanted out.

That fall, a group of delegates gathered in Santa Barbara to discuss the division of California. In their report, the delegates—primarily Hispanic and from the South—stated: "It is a plain truth that whatever of good the experiment of a state government may have otherwise led to in California, for us, the southern counties, it has proved only a splendid failure." They recommended dissolving the marriage of North and South California, which they claimed was "in contradiction to the eternal ordinances of nature." They asserted that nature had "marked with an unerring hand the natural bounds between the great gold regions of the northern and internal sections of the state and the rich and agricultural valleys of the south."[18] Although nothing came of this convention and report, they did serve to publicize the Californios' serious concerns about the new state government.

Dissolve the Unholy Union!

Throughout the 1850s, the subject of dividing California came up again and again. Each session of the legislature debated its merits and potential pitfalls. Representatives from Southern California advocated separating their fate from that of the Yankees in the North. From 1852 on, even the governor showed sympathy for their cause.

In a speech before the legislature, Governor John McDougall stated that the Southerners, mainly farmers and cattle ranchers, were paying an inordinate share of state taxes and were not seeing the fruits of their contributions. With a population of about 6000, the six southern counties paid, in 1851, $42,700 in taxes, he noted. The mining counties in the North, on the other hand, were getting much more for their money. The twelve mining counties, with 120,000 people, paid only $21,253 the same year. And the North had three times as many representatives in the legislature as did the South. The governor recommended calling a convention to amend the constitution and talk about splitting the state.

For the first time, a rural-urban rivalry can be detected in the state's interregional relations. The Californios resented the Yankee urbanites of San Francisco and their commercial ambitions. Urging a political separation from the more populous North, southern delegates sought to include "only such agricultural and grazing counties as are identified with us in interest."[19] They did not want to have to compete with the wealthier, dominant counties with whom they shared no affinity.

The natural distinctions between urban and rural lifestyles and values continue to this day. They have themselves sparked split-state drives in the past. Today they tend to cloud the issues raised by modern advocates of a North-South split.

During the 1854 legislative session, an assemblyman from San Bernardino County, Jefferson Hunt, introduced a bill to create a "State of Columbia" out of the southern counties of California. Instead, the legislature chose a plan to trisect the state. Several representatives argued that residents in the far northern counties had as little representation in the US Congress as did the southern counties. Both rural areas and sparsely populated, the far north and the South felt neglected by the better represented urban center of San Francisco, they argued. This group favored the trisect plan because it would give the Pacific Coast four more senators in Washington.

The plan would have divided California as follows: The counties as far north as Monterey, Merced, and part of Mariposa would become the "State of Colorado"; the far northern counties of Klamath (now Modoc), Siskiyou, Humboldt, Shasta, Trinity, Plumas, and portions of Butte, Colusa, and Mendocino counties would become the "State of Shasta"; and the rest of the territory would constitute the "State of California."

The bill passed the assembly, but the session ended before the senate could vote on it. The next legislature was expected to pass the three-state proposal. It probably would have, but for "one of those political cataclysms that occasionally overwhelm the schemes of politicians," as historian J.M. Guinn put it.[20]

In 1856, the California Democratic Party was undergoing an internal tug of war. Led by none other than the indomitable US Senator William Gwin, the Chivalrists supported the southern US states and their institutions, especially slavery. At the other end of the political spectrum was the liberal faction, led by California State Senator David C. Broderick, the son of Irish immigrants and himself a stonecutter. Broderick and the liberals attempted to wrest control of the party from the Chivalrists. The feud between Gwin and Broderick, who later also became a US Senator, distracted the attention of the legislature that year, muting the discussion of state division. But the state's newspapers, as they had throughout the debates, kept the subject alive.

In 1853, the *Sacramento Union* ran an editorial stating that, "A division of the state into two or more states is a political necessity which will be recognized by all parties sooner or later."

Taking a stand altogether different from most northern papers, the *Union* denounced the assertions that separation efforts were motivated by those who would bring slaves to California. The paper labeled such charges "the production of a fevered fancy, an imagination so diseased upon the subject of slavery as to be unable to view the subject [of division] through any other than a distorted medium."[21]

The *Daily Alta California*, an influential northern newspaper, editorialized against a division of the state, charging that slavery propagandists were behind it. The *Los Angeles Star* quickly responded with a defense of division. The southern paper stated that the desire for a separate state was the natural outgrowth of the sorry situation in Southern California. It added that

Southern Californians greatly resented the charges that the plan was slavery-inspired. Such accusations obscured the real issues, the paper said.

Although it is true that Gwin and the Chivalrists were dying to see slavery established in California and tried every tactic they could think of to see it happen, this was not the intent of the Californios who wanted division. The rural, Hispanic Southerners had legitimate complaints, but the Chivalrists' behind-the-scenes plots aroused suspicions about the motives of *any* split-state advocates, tarring them all with the pro-slavery brush.

The slavery charges served to dissuade support for a split and helped to keep the state whole. This is not unusual in California; obfuscation of the real issues is as much a part of the American political process as is the electoral college, and it has plagued split-state politics in particular all the way up to the present.

The Pico Act

The decade-long movement to divide California reached its climax in 1859. Dismayed by the Yankees' attempts to deprive the Californios of their land, Don Andres Pico decided to force the issue. Pico, the Californio hero of San Pasqual and now himself a prosperous ranchero, owned thousands of acres in Southern California with his brother Pio. Andres had been elected to the assembly from a district encompassing Los Angeles, San Bernardino, and San Diego counties. At the beginning of the legislative session of 1859, Pico introduced a resolution to divide the state. The reasons he gave resembled those of earlier divisionists: the area of the state was too large, the Californios were overtaxed and underrepresented, and uniform legislation was unfair to the South. His bill was referred to a special committee.

While the committee pondered the bill, another was introduced in the assembly. Representatives from the far northern counties wanted to create a "State of Klamath." They too felt underrepresented, overburdened, and neglected by the state government. Historians disagree as to the real motive behind the "State of Klamath" proposal; some believe it was offered to ensure that the far northerners' interests would not be forgotten while the legislature considered Pico's proposal.

Although the "State of Klamath" never materialized, the proposal set a precedent for later moves by California's far North to detach itself from the rest of the state.

In a few weeks, the committee finished its deliberations on the Pico bill and reported back to the legislature. The members were not at all certain that Pico's proposal was legal under California's constitution, but they whole-

Andrés Pico, victor at San Pasqual and author of split-the-state
legislation. Courtesy, the Bancroft Library

heartedly embraced the idea of dividing the state. Refining Pico's plan, they
came up with a bill of their own. Theirs would create the "Territory of Colorado" out of the six counties below the thirty-sixth parallel (a few miles south
of Big Sur).

To make things interesting and more democratic, the bill required not only
approval of both houses of the state legislature, but also a vote of the people
living in the area to be segregated. If two-thirds of the voters cast their ballots
in favor of territorial status, the measure would pass. Then it would face the
test of national acceptance; both Congress and the President would have to
approve it before it could become law.

While the delegates debated the issue in the legislature, newspapers across
the state buzzed with arguments pro and con. The *Sacramento Union* acknowledged the sincerity of the South's cause and stated: "The members of the south
in the convention to form a state constitution desired to be left out, but as they
were informed that great advantage would result to those counties, they willingly
submitted. A ten years' experience has convinced them they were deceived."[22]

On March 25, the assembly passed the bill, 33 to 25; the California senate
approved it on April 14—15 to 12—and the governor signed it five days later.
The vote broke down along sharply defined regional lines. All of the delegates
from Southern California voted for it; those from the North were split: eleven

northern senators voted for it and twelve against. In the assembly, 27 northern representatives favored it and 24 voted against the Pico bill.

As the act required, the question "For a Territory or Against a Territory" was put to the people in the southern counties. They voted as follows:

	For	Against
Los Angeles	1407	441
San Bernardino	421	29
San Diego	207	24
Santa Barbara	395	51
San Luis Obispo	10	283
Tulare (now Kern)	17	—
Totals	2457	828

Plainly more than two-thirds of Southern California residents wanted a divorce from their neighbors to the north. Three out of four voters south of the proposed border voted to secede.

Now it was up to Congress to decide California's fate. Governor Milton Latham, a northern man with southern principles and an avowed divisionist, had just been elected to the US Senate—to replace Senator David Broderick, who had been killed in a duel. As governor, Latham sent the results of the popular and legislative votes to Washington. As senator-elect, he knew he would probably have to vote on the issue in Congress. This placed him in a ticklish situation, but he decided to follow his convictions. In a long letter to President James Buchanan, Governor Latham explained why he thought Congress should approve the division plan:

"The origin of this act is to be found in the dissatisfaction of the mass of people, in the southern counties of the State, with the expenses of a State government. They are an agricultural people, thinly scattered over a large extent of country. They complain that the taxes upon their land and cattle are ruinous—entirely disproportioned to the taxes collected in the mining region . . . .and that there is no remedy, save in a separation from the other portion of the State."[23]

Latham asserted that a division of California was entirely legal. Conceding that the act had not been subjected to a vote of the whole state's electorate, as required by California's constitution, he argued that the US Constitution's requirements for creating states* took precedence over the state's laws. He stated that the California legislature had met the federal stipulations and added that nothing in the US Constitution prevented a severed portion of a state from choosing a territorial instead of a state form of government. Latham's well-reasoned arguments would be used by separation advocates and foes alike in the years to come. They are still referred to today (see chapter 5).

Congress was in no mood to hear about sibling rivalries on the far frontier. More serious business was at hand. The Southern states of the Union were threatening to secede. The Pico Act never came to a vote in Washington. The Civil War broke out the year after it was introduced and California's request for separation was forgotten.

Although California was across a continent from the battlefields, the Civil War sparked violent sectional conflicts and inflamed the hearts of people on the Pacific Coast, most of whom had migrated west from one of the warring states. About three-eighths of all Californians supported the Confederacy during the Civil War. Several newspapers, the Chivalrists, and a number of communities—Los Angeles, Visalia, El Monte, Ventura, and San Bernardino— were among the most blatant Southern sympathizers.

"Paris of the West"

After the war, split-state agitation all but died out. Having come so close in 1859 but failing in the end, separationists were glum about the prospects for success. And since many were rancheros, their attention turned to defending their land claims and trying to make ends meet. Besides, it looked as if the Yankees were settling in—in all parts of the state. If these ethnocentric, determined people did not want the state divided, there probably wasn't much the Californios could do about it, after all.

In 1869 the transcontinental railroad was completed. In an elaborate ceremony at Promontory Point, Utah, California Governor Leland Stanford, one of the backers of the railroad, drove in the golden spike that joined the rails of the Central Pacific Company with those of the Union Pacific. The act marked the end of the State of California's 20-year isolation and irrevocably tied its destiny to that of the rest of the United States.

---

*"New States may be admitted by the Congress into this Union; but no new State shall be formed or erected within the jurisdiction of any other State; nor may States be formed by the junction of two or more States, or parts of States, without the consent of the Legislatures of the States concerned, as well as of the Congress."   —Article IV, Section 3, US Constitution

Now that travel from the East Coast and the Midwest to San Francisco was only a matter of a few days' train ride, some 700,000 passengers a year began to make their way to the West Coast.

In addition to more settlers, the railroad brought prosperity to California. The state's biggest and most important city, San Francisco, was soon transformed into the distribution center for the entire West Coast, from Oregon and Washington to Baja California. During the 1860s, the city's first merchants established themselves along the bay. Peter Wiley and Bob Gottlieb point out in *Empires in the Sun* that Domingo Ghirardelli, the chocolate maker, arrived from Italy; from Germany came Levi Strauss, the pants maker, Anthony Zellerbach, the papermaker, and Isaac Magnin, the shopkeeper. The prodigious Comstock Lode, just across the state border in Nevada, brought the "Bonanza Kings"—Fair, Flood, O'Brien, and MacKay—to San Francisco. There they met George Hearst, another Comstock millionaire, whose heirs would turn his mining fortune into a vast communications empire. Henry Wells and William Fargo founded Wells Fargo and Company, the banking and transportation firm. Soon factories, canneries, and packinghouses sprang up along the waterfront from North Beach to Potrero Point, making it into the largest concentration of industry on the West Coast.

San Francisco was also home to "the Big Four," the men who had built the transcontinental railroad. Former merchants who had come to California during the Gold Rush, Leland Stanford, Charles Crocker, Mark Hopkins, and Collis Huntington formed the Central Pacific Railroad in 1863 and won passage of federal funds to build the rails from California to Utah. Since the West had been largely unpopulated until then, the Big Four had convinced Congress to grant them generous strips of land alongside the tracks. By the time Stanford drove the golden spike, the Big Four had earned a formidable reputation. They had bought out most of the major railways in the state—including those of the Southern Pacific Company, whose name they also acquired—and most of the shipping routes as well. Southern Pacific became the largest private landowner in California, with the political clout to match. The company's monopoly on the state's transportation systems, its refusal to lower rates, and its untoward influence over the courts and the legislature won it the nickname "the Octopus" and raised the ire of Californians in every section of the state.

While San Francisco was blossoming into a cosmopolitan center and earning itself the nickname "Paris of the West," Southern California was being discovered for the first time by Americans. Attracted by the superior climate (or

The Big Four: (l-r) Hopkins, Huntington, Stanford, and Crocker.

repelled by the North's obsession with gold and money), many of the new-comers on the transcontinental rails headed south upon arrival. The railroads launched a major public relations drive to "sell California," which, to a large extent, meant the weather of Southern California.

In anticipation of the expected boom, the Southland prepared itself. "Pictur-esque cottages were torn down in Santa Barbara, Los Angeles, and San Diego to make way for the new buildings in the cities-to-be. Wharfs, railway termi-nals, hotels, warehouses, and churches sprang up," wrote Carey McWilliams. By 1876, the sleepy cow town of Los Angeles had undergone a remarkable transformation: "The little town of 6000 population suddenly began to think of itself as another San Francisco."[24]

Although thousands of tourists were lured to Southern California by descrip-tions of it as "a Mediterranean land without the marshes and malaria," "the American Italy," and "the new Greece," the big boom of the '70s never quite materialized. Most visitors went home. Those who did stay, however, were very

Courtesy, the Bancroft Library

different from the people who had settled in the North years before. Instead of golddiggers, farmers and their families settled in the Southland and founded the communities of Riverside, Ontario, and Santa Anita. The differences between the North and the South continued to grow.

Economics Divide the State

In 1877, a Los Angeles judge tried to revive the Act of 1859 and split the state. Judge Robert M. Widney wrote in the *Los Angeles Express* that the industries of the two sections of California were entirely distinct and that northern control of the corporations hampered southern progress. If the state were divided, he argued, the South would receive the funds it so badly needed to improve the harbor in Los Angeles and build more railroad facilities. At the same time, it could install a more honest and economical state government. His arguments echoed the anger and frustration of other Californians who were demanding the state's constitution be amended to knock the big corporations—particularly "the Octopus"—down to size. The legislature and the governor acceded to these demands. The Second Constitutional Convention was set for September 1878.

One hundred fifty-two delegates from all over California gathered in Monterey this time. From the beginning, it was clear that the varied industrial and commercial interests of California did not see eye to eye. The farmers opposed the miners because of water rights; small farmers opposed big farmers who were allied with the big corporations; the merchants were angry with the banks and the railroads for trying to overcharge them; and the labor party opposed them all.

Each group wanted its complaints answered by a constitutional amendment. The universal theme of the convention was opposition to "big business" and "big agriculture." The preamble promised "to wrest the government from the hands of the rich and place it in the hands of the poor." The delegates voted to set up a commission to regulate railroad fares (heretofore successfully thwarted by Southern Pacific); to create the State Board of Equalization to ensure that property tax assessments were equalized among all counties; to revise the tax laws; to ban "special interest legislation," a term for laws favoring the big corporations and landowners; to define irrigation water use (though not to amend the state's troublesome water rights code); to make education compulsory; to approve the state's first eight-hour work day; and to require that all court actions be conducted in English, not Spanish.

This last action, more than any other taken in Monterey in 1878, signaled the complete Americanization of California.

After the convention ended, the legislature returned to its work. Dominated by representatives from the more prosperous North, it passed laws creating the San Francisco Mining Exchange (the first stock exchange on the West Coast) and establishing an agricultural school near Sacramento, but it voted down funding for improving the harbor in Los Angeles and for a branch of the state teachers school, known as the "normal school," in the Southland.

Southern Californians had for years been trying to establish a statewide irrigation system with designated irrigation districts. Farmers—primarily residents of the southern and central valleys of California—felt such legislation was crucial to the development of irrigated agriculture in the state. Adding insult to injury, the legislature rejected all the proposed irrigation district bills, calling them "special interest legislation." Southern delegates also wanted the state's water laws amended to accommodate the farmers' needs for irrigation water from streams not crossing their land. Farmers insisted that the existing doctrine of "riparian rights" was not suited to their needs, but to those of Northern California miners. These complaints, along with the North's refusal to respond satisfactorily to the Southerners' demands, set the stage for California's first regional water war, as well as impassioned pleas to "split the state!"

Gold vs. Grain: The First Water War

There was no doubt that the northern part of California controlled the legislature in the middle-to-late 1800s. The miners who had come to make a fortune had either done so and gone home or they had decided to stick around to spend or invest it. Others had joined together to form large mining companies to try to extract the last remaining gold particles from the Sierra foothills. These wealthy miners and entrepreneurs lived in the North, most of them in San Francisco. It was easy for them to lean on their pals in the legislature for

Placer mining with the aid of a pan.
From *California, a History of the Golden State*

favors. The capital was, after all, in Sacramento, only a matter of hours from San Francisco, but it was at least a day's journey from Southern California.

The constitutional amendments passed at the Second Constitutional Convention failed to bring about the promised reforms. The commission set up to regulate railroad rates had a pro-railroad majority, and the courts always seemed to side with the wealthy railroads in legal disputes. More than a few of those disputes involved farmers who wanted the right to bring water from streams, across railroad-owned land, to irrigate their fields. The courts ruled that riparian rights—those belonging to landowners with land adjacent to streams—took precedence over "appropriative rights"—those of the would-be irrigators who did not own stream-side land. This policy left California's farmers high and dry. They wanted the legislature and the courts to come to terms with their needs to appropriate water; but their requests fell on deaf ears.

Even more frustrated than the would-be irrigators around the state were farmers in the Sacramento Valley whose land was periodically flooded as a result of upstream mining operations. After the placer miners had scoured as much gold and gravel as possible from Northern California streams, the large mining companies devised a technique to get at the veins buried deep within the hillsides. "Hydraulic mining" used pressurized water to wash the earth from gold-bearing hills. Water from high mountain lakes, reservoirs, and streams was diverted down the mountains in flumes, then channeled into huge

Hydraulic mining, introduced in 1852, abolished in the 1880's

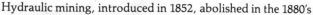

iron nozzles, called "monitors." Miners directed the spray at the chosen hillside and watched as it stripped away the earth, freeing the bits of gold from their ancient captivity.

The monitors' power was tremendous. It was not unknown for a man or an animal, unlucky enough to get in the way, to be killed by a direct shot of the pressurized water from as far away as two hundred feet. One can imagine what a hillside looked like after a day's "washing."

The dirt was then channeled into a system of sluices so that the miners could extract the gold particles. After the miners had reduced the mountain to a molehill, they sent the resulting mud and gravel downstream, via the over-burdened riverbed.

Obviously, the farmers downstream were furious. After years of watching helplessly as the mining debris washed over their land and ruined their crops, farmers in Butte County had had enough. They weren't going to take it anymore. Pooling their resources, the farmers filed suit, in peach farmer A. J. Crum's name, against the Spring Valley Mining Company. They demanded $2000 in damages and an injunction to halt the mine's operations. The farmers of Butte County lost their suit but they instigated a violent, eight-year battle between the hydraulickers and the downstream farmers.

Within a few years, the battle had moved to the legislature. Moderates, including William H. Parks, a big, somber, bearded farmer from Sutter County, urged passage of a bill to solve the debris problem by constructing a series of dams and levees to restrain the devastating flows. Farmers who had lost their land to the miners' waste products demanded stronger action; they urged an outright ban on the destructive enterprise of hydraulic mining.

One, George Ohleyer, had farmed in the Sacramento Valley since 1852. In 1862, the banks of the Yuba River overflowed and covered his fruit orchard with a thick layer of sand and mud. He was forced to sell. He bought another parcel and started over. Eventually he became editor of the *Yuba City Farmer* and an influential molder of public opinion. Citing a grim list of the thousands of dollars in damage done and the thousands of acres destroyed by mining debris, Ohleyer told a meeting of farmers in 1874 that the only solution was a complete ban on hydraulic mining.

The legislature, as was its wont, took the middle road. After all, mining had been there first, and it had been the state's principal industry for many years. In 1880, California's lawmakers adopted a bill requiring construction of a system of levees on the Bear, Feather, and Yuba rivers to prevent debris from choking downstream waterways and flooding farmlands. The state would foot the bill.

Ohleyer and the anti-hydraulickers were incensed. Not only had the legislature failed to ban the hated industry, but its bill would actually serve as a subsidy for the miners! Instead of placing the economic burden for constructing the levees on the responsible parties—the mining companies—the Drainage Act would levy a tax on every property owner in California.

Landowners all over California were livid. The hydraulic mines were "too openly the product of wealthy capitalists for people to submissively pay even small taxes for their benefit," wrote Robert Kelley in *Gold vs. Grain.* "The Drainage Act smacked of special interest legislation in a time when this term meant corruption."[25]

Southern Californians in particular expressed outrage at the new property tax. They felt that the Sacramento Valley should take care of itself, that the miners should pay all the costs of constructing the levees, and that the miners should also reimburse the farmers whose land had been ruined.

The day before the bill passed the legislature, an editorial appeared in the *Los Angeles Express,* stating, "The Mining Debris bill . . . is neither more nor less than an impudent raid upon the taxpayers of the state . . . . there is neither sense nor justice in levying that tax upon Southern California which has neither lot in the benefits nor injuries arising from hydraulic mining."

The themes are familiar. We saw something very similar in the 1980–82 debates on the Peripheral Canal, not to mention the fights over passage of the State Water Project in the Burns-Porter Act of 1960. Regionalism is a strong force in California politics.

"Division Is a Necessity"

The vitriolic *Express* writer was at it again, three days after the Drainage Act passed: "The division of the State of California and the erection of a new Commonwealth out of the southern portion is becoming more and more exigent every day. The diversity of interests of the two sections is made conspicuously apparent by the action of every Legislature . . . . Small as our population is, the rest of the state has persistently refused to give us even a just representation for that population . . . . If we ask for any concession to which we believe our section is entitled, we are insultingly scouted; if we protest against any oppression, we are laughed at. If we object to the concentration of all the public institutions in a few and favored counties, and the distribution of all the revenues of the State among those counties, we are treated as soreheads."[26]

Such rhetoric inflamed the long-suffering Southerners. They persuaded

Assemblyman Del Valle of Los Angeles to introduce legislation to divide the state in two. But the session ended before action could be taken.

Divisionist sentiments seethed. A mass meeting in Los Angeles in early 1881 broke up with "three cheers for the State of Southern California!" The southern press continued to rail against the selfishness and corruption of the North. The *Los Angeles Herald* even came up with a cockamamie proposal to combine Los Angeles, San Diego, and San Bernardino counties with the Territory of Arizona to create "Calizonia."

While Los Angeles papers leveled venomous attacks on the Drainage Act and its hated tax, the northern press defended the act and the principle of general taxation. "Nations nor States could be formed if the theory of strictly local taxation were adhered to," went a January 1881 *Sacramento Union* editorial. "It is the principle of mutual aid which the nation has long practised and which has always had the best results."[27] As the drive to repeal the Drainage Act came to a head in the legislature, the Sacramento paper argued that without the law and its necessary dams and levees, the Sacramento Delta waterways would become permanently unnavigable. Voicing an argument similar to one that would be leveled in favor of another piece of legislation 100 years later, the *Union* stated, "If the Sacramento Valley goes, the harbor of San Francisco will go too."

Southern California separationists called a convention for later in 1881 to discuss how to go about splitting the state. In spite of all the emotional and self-righteous mud-slinging going on, not too many delegates made it to the Los Angeles meeting. One historian speculated that because many Southerners resented Los Angeles and the Southland's largest city's desire to be the capital of the new state, the other counties did not send as many delegates as they could have.

The disappointed delegates to the 1881 convention decided to wait "until the population of the new district was large enough to insure its reception as a state," before taking any action, as Charles Willard stated.[28]

Meanwhile, opponents of the Drainage Act got nowhere in Sacramento. So a group of farmers and local businessmen from the Sacramento Valley tried another tack. They filed a lawsuit challenging the legality of the act. In 1881 the California Supreme Court threw out the Drainage Act, calling it an unconstitutional assumption by the state of an essentially private concern. The legislature, the justices ruled, did not have the authority to tax everyone to benefit a few, just as the angry Southerners had argued.

Three years later, a federal circuit court resolved the impasse. It issued a perpetual injunction against the discharge of hydraulic mining debris into California rivers. The judge ruled that because debris created irremediable and uncontrollable damage to the community, the general welfare required an end to such discharges. Thus, an entire industry was shut down in one of the nation's first environmentally responsible decisions. In addition, California agriculture had won one of its first major victories in the courts. The state's oldest and leading industry was being shoved aside. The challengers, Central Valley and Southern California farmers, would now take the place of miners, at the top.

The South Booms

By this time, Southern Californians had become thoroughly enamored of the topic of dividing the state. Businessmen and workers alike would spend hours discussing the whys, wherefores, and hows of permanently breaking it off with the dastardly North. The southern press continued to run vituperative attacks on the greedy corporations and corrupt politicians in San Francisco and Sacramento. In 1885, the State Board of Equalization, created seven years earlier to prevent oppressive taxation, raised the assessed valuation of Los Angeles by $5 million. Southerners, up in arms, demanded an immediate divorce. But their cries went unheeded.

The next year, the Santa Fe Railroad completed its line to Southern California and challenged the powerful Southern Pacific's monopoly on the southern routes for freight and, more important, tourists and immigrants. A ferocious rate war ensued, giving birth to the "Great Boom of the 1880s."

Because they came by train, these newcomers were a different breed from California's early settlers. Most were well-to-do, the families of merchants, doctors, pharmacists, and others who could easily afford the price of land when they arrived. The railroad offered cut-rate fares from the East Coast and the Midwest, and provided a week's free lodging to prospective property buyers. Lured by the low rates, the fine climate, the promise of a new life, and the fertile soils of the Southland, they rushed to buy their tickets and headed west on the first train they could get.

"Instead of *Shank's Mare* or prairie schooner, or reeking steerage," wrote historian Charles Fletcher Lummis, the '80s settlers "came on palatial trains; instead of cabins they put up beautiful homes; instead of gophering for gold, they planted gold—and it came up in ten-fold harvest."[29]

The differences in motivation and background between the settlers who made their homes in Southern California in the 1880s and those who had ventured to Northern California 30 years before deepened the gap between the two parts of the state. Previously dominated by the Californios' traditions and customs, the cultural landscape of the Southland was changing. Along with the well-to-do midwestern families came their conservative political and religious values. Eventually, such values would clash with the "devil-may-care" philosophy of the many Northern Californians who lived recklessly and lawlessly in the goldfields and in "Baghdad-by-the-Bay." But the confrontation did not come for many years.

Southern California's population leaped from 64,371 in 1880 to 201,352 by 1890. The boom brought long-sought and lasting improvements to the Southland. In anticipation of the hordes, developers paved roads and sidewalks where only dirt paths had been. A reliable system for domestic water service was established, as were irrigation companies, streetcars, and several educational institutions—including the University of Southern California and Occidental College. Finally, after years of fruitless attempts by southern legislators, a branch of the state normal school (now the University of California at Los Angeles) was built in the South.

At the end of the legislative session of 1888, General William Vandever, a representative from Ventura, introduced a bill in the assembly to create the "State of Southern California." No doubt cocky from the region's recent boom, Vandever proposed that the new state extend farther north than any previous split-state plan—to embrace Alpine, Tuolumne, Merced, San Benito, and Monterey counties. The bill never left committee. Southern Californians were absorbed in instituting long-needed services and building long-awaited facilities.

In 1906, disaster struck San Francisco. The devastating earthquake and fires that swept the city crippled its social and economic structure for many years. An industrial giant before the quake, San Francisco now barely managed to crank out a can of tuna. In contrast, Los Angeles' economy fluorished; manufacturing in the southern city registered a 100 percent increase in the first decade of the new century.

Los Angeles' growing industrial presence was largely the product of one determined man, General Harrison Gray Otis. Since San Francisco had, by 1890, a 40-year head start, Otis, publisher of the influential *Los Angeles Times,* and others of a similar mind, realized that the only way to establish Los Angeles as an industrial center was to undercut the high wage structure of San Francisco, a strong union town. The low wages would attract capital away from the North. Taking advantage of the surplus labor in the region—the steady influx of settlers—General Otis thwarted unionization efforts in Los Angeles for years, keeping wages in the southern city 30 to 40 percent lower than in San Francisco and attracting crucial investment funds to the South.

In the first decade of the twentieth century, there were three moves to split the state in two. The first one, in 1907, was the result of the North-South industrial feud. The second, in 1909, was a knee-jerk response to increased taxes in the South. The third was another half-hearted attempt by the far northern counties to form a separate state, this time to be called the "State of Siskiyou."

Of the three, the most important was inspired by Senator Robert N. Bulla. In a 1907 speech before the Los Angeles Sunset Club, Bulla addressed three questions: Could, should, and would the state be divided North-South? He argued that the Act of 1859 was still in force and that only the consent of Congress was needed to accomplish a separation. He stated that California should be divided because the trip to Sacramento was too long and too expensive for Southerners; because the South wanted a separate government; and, best of all, because Californians would get two new representatives in Congress. In response to concerns that a new bureaucracy would be costly, Bulla claimed

O'Farrell from Powell Street, San Francisco, 1906. Courtesy, the Bancroft Library

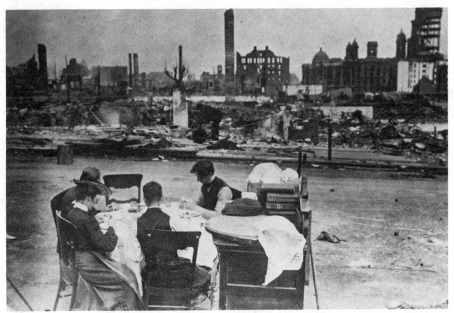

Dinner as usual, San Francisco, 1906. Courtesy the Bancroft Library

that Southern Californians could run their own government more efficiently than Sacramento bureaucrats and officials did then. So it wouldn't be long before the higher taxes required to finance a division would be returned to the people in the way of services and lower taxes in the long run. He urged the legislature to revive the Act of 1859 and create a new "State of Los Angeles."

Bulla acknowledged that at least one major obstacle existed. The boundary line set in the Act of 1859 would cut Los Angeles off from Inyo County. By this time, the South's burgeoning metropolis had reached out to the Owens Valley for water. Largely because of this, Bulla's plan to resuscitate the Pico Act failed.

Southern Californians' enthusiasm for splitting the state—so strong for so many years—was finally beginning to subside. An opponent of Bulla's proposal, Senator H. E. Carter, wrote in the *Grizzly Bear*, the publication of the Native Sons of the Golden West, that the sectional differences between the North and the South were disappearing. He argued that Southern California was starting to see the benefits of its healthy contributions to the state treasury. He added: "Contrary to being unable to get needed legislation, the Southern California delegation, for the past eight years, has been able to and did get through all and every bit of legislation requested by the people of Southern California."[30]

While this last statement was no doubt hyperbole, it does reflect the changing perception Southerners had of themselves and their position in the state.

The End of an Era

Up to this point, the Southland had been the neglected, overburdened "very small tail to a very large kite," as the *Los Angeles Express* wrote in 1880. The desire to sever the state along the Tehachapis (or any other east-west line) had in every instance (save the far north's semi-serious efforts) originated in the South. Southerners resented paying a disproportionate share of state taxes, while the golddiggers and wealthy entrepreneurs of the North enjoyed greater representation in the legislature and the favor of the courts.

Intensifying the Californios' wish to be rid of the unfair taxes was their fear of the Yankee newcomers, so aggressive and insensitive to their traditions and culture. Thus, both objective motivations (taxes) and subjective ones (fear) contributed to these early split-state movements. Throughout the history of the Golden State, split-state drives have followed a similar pattern.

In the same vein, resistance to state division—traditionally stronger in the dominant, more populous part—has been based on both rational and irrational concerns. On the irrational side are usually fears that disaster will strike if a state is suddenly torn asunder. In the 1850s, this manifested itself in the pervasive—and largely unjustified—fear that slavery would be instituted on the West Coast if the state were split in two.

On the rational side, there are several good, sound reasons to oppose a split-state. It was, and is, an extremely expensive and complicated proposition to divide a state and set up a brand new government. Politically it is next to impossible to accomplish: both the US and California constitutions have thrown up stumbling blocks that increase in volume the closer one gets to success (see chapter 5). Arguably the most challenging obstacle is a vote of the legislature and of the people of the entire state. History and common sense indicate that the section holding the reins at any given time has no incentive to drop them in the name of "equity" or "justice."

Around the time of World War I, the situation would abruptly shift. The South would grow. Los Angeles would consolidate its power, and the North would reconsider its partnership in the Golden State. Those who were formerly powerful, now feeling impotent, would start making noises about seceding. And former underdogs—now on top—would be the ones extolling the virtues of union.

2

Los Angeles Rising
1914–1982

Ever since the discovery of gold, a pendulum of power has swung freely over the state of California. During the early days of statehood, the North boomed and bustled, while the South, for a time a dusty, forgotten outpost, cried out for recognition. The people of the North must have thought their ascendancy would never be challenged. The pendulum swung as far as it could toward the North, however, and then it began to swing back. The lonely cow counties of the South grew in discrete bursts, in staggered waves of immigration. By the year 1910, the shaft of the pendulum was nearly centered. The population of Los Angeles had climbed from about 100,000 in 1900 to 319,000 in 1910. San Francisco was growing, but much more gradually. By the second decade of the century, the two cities were nearly equal in size.

There was a brief lull in intersectional rivalry during these years. Some observers seemed to sense the comfortable feeling of relative parity between the two regions. "The occupations and character of the people are coming more and more into harmony," wrote Joseph Hayford Quire in 1910. "One race of people exists where two had formerly lived. All conditions go to show that we will have no 'North California,' 'Central California,' or 'South California,' but instead a unified, a strong, and an incomparable Golden State of California."[1]

This "eye-of-the-hurricane" period did not last for long. The pendulum continued its inexorable swing to the

> **"We of Southern California are foreigners to Northern California. We are two separate countries."**
>
> **—Max Miller**

45

south. By 1914, the effect had become noticeable to Northerners. A great influx of midwestern farm people had come to Los Angeles, and their conservative values clashed with those of the freewheeling San Franciscans. The sharpest issue in the elections of 1914 was Prohibition.

Several western states passed Prohibition initiatives in 1914, but few people expected California to join the movement. The vote was startlingly close, a tribute to the growing political clout of the conservative South. *Sunset* magazine declared the vote a narrow victory for "California's wine industry and San Francisco's robust thirst." For the hard drinkers and the liberals in the North, the implications of the elections of 1914 were clear: the pendulum was moving south.

When the pendulum of power swings away from a region in California, the cry of secession begins to arise there. In anticipation of the 1916 elections, San Francisco engineer Russell Dunn and others formed the People's Association for Changing the Boundary of California by Amending the Constitution. Dunn listed his group's reasons for advocating a split in the *Los Angeles Express:* first, the eastern and midwestern immigrants in the South were bent on forcing their will on the Northerners; second, southern plans were discouraging out-of-state investors; third, mining interests were being hurt by Southland-supported laws; fourth, the southern press for Prohibition was intolerable to the North; and fifth, the North needed the increased congressional representation that a split could provide.

Dunn's group proposed splitting the state at the Tehachapis (San Diego, Riverside, Orange, Santa Barbara, Ventura, Los Angeles, and San Bernardino counties composing the southern state). At one point, Dunn offered publicly to trade Inyo County and its water to the South for Santa Barbara, the only southern locale he deemed worth keeping. Dunn's movement caused a mild stir but then faded into obscurity, dying perhaps, as the *Los Angeles Times* noted wryly, "by the weight of its own name."

With each succeeding election, the antagonism between North and South increased. One letter-writer to the *San Francisco Chronicle* abused the new Southerners as unwelcome migrants from the "crude, provincial regions of the Middle West. . . . I notice in the election returns," he continued, "that the people of the sanitary southland are preparing another slaughter of real Californians. . . . Give 'em a separate State and let them call it Puritangeles."[2]

This gentleman and other San Franciscans must have been deeply affronted when a Prohibition referendum was finally passed in 1922, by a margin of

445,000 to 411,000. The shaft of the power pendulum had passed the mid-point of its arc.

Jes' Grew

At the turn of the twentieth century, Los Angeles had been a modest, second-rate town. The 1910 edition of the *Standard Dictionary of Facts* described it simply as: "Los Angeles—on the Los Angeles River, 480 miles south of San Francisco." By the 1930s, however, Los Angeles County included more than 2 million people. In the four decades between 1900 and 1940, the population of Southern California increased 1535.7 percent! This amazing metamorphosis

Los Angeles Times, 1926, Ted Gale

put unprecedented stress and strain on California's political unity. In fact, it is impossible to understand California's North-South feud without a grasp of the mechanics of the Southland's incredible growth.

To understand what happened to California, try to imagine an analogous growth and shift of power in another state. What if Rochester, New York, suddenly became a metropolis of 15 million people? Suppose Scottsdale, Arizona, outgrew Phoenix in a matter of decades. Imagine the consequences if Springfield, Massachusetts, were to become larger than Boston. The result for California was a nasty set of growing pains, problems the state has yet to overcome.

Those planners who have tried to keep up with California's growth have had a hard time. Particularly in the Southland, the area keeps outgrowing the solutions. Today's problems with water politics reflect this situation. From a Northern California point of view, the problem with water exportation is not so much the continued permission of present exports, but the implication of many more decades of explosive Southland growth. Can any amount of water hope to slake the thirst of this burgeoning beast? a northern environmentalist might think. From the viewpoint of a southern planner, the problem is similar. The danger lies not with the inadequacy of current water supplies, but with the unknown factor of future growth in the area.

The rapid growth of Los Angeles and its environs, then, lies at the heart of California's intrastate friction. How did this growth occur? Was it man-made or accidental? Was Los Angeles a place that, like Topsy, "jes' grew"? Actually, both human and natural causes contributed to the phenomenon. The best place to start an explanation is with the weather.

Climate

Southern California sunshine probably deserves the ranking of the first cause. Even a die-hard Northerner would probably admit that the weather in Southern California, in the absence of acrid photochemical smog, is just about perfect. Mountain, desert, and seashore converge in the LA Basin. The result is a happy blend of the best of all three worlds: warm, sunny days, cool nights, fresh breezes, light rainfall, and bright, snow-capped mountains on the horizon. It's the sort of climate that turns businessmen into body builders and housewives into marathoners. It calls the tired, huddled eastern and midwestern masses to throw away their galoshes and overcoats and umbrellas, to forswear their pasty complexions, to come to Southern California and try to live forever. It is, one might say, a sort of instant paradise: just add water.

Water

The shortage of H_2O, the world's most common compound, has always been Southern California's most intractable problem. Apparently, Mother Nature gives us all those terrible winter storms and hibernal discomforts for a reason. Snow and rain and cold and slush bring us rushing rivers and booming springs, which in turn enable us to garden, drink, make coffee, give parties, and have plumbing throughout the year. Los Angeles is in a classic have-your-cake-and-eat-it-too situation. The very climate that attracted so many could by definition support only a few, unless...the life-giving water could be imported (stolen if need be) from some less populated place. Starting in 1905, officials of the City of Los Angeles began to buy up land in the Owens Valley, more than 200 miles away. The people of the city, the population of which had more than doubled in five years, approved a bond issue to build a delivery system to bring Owens River water to LA. This aqueduct was completed in 1913. It delivered four times as much water as the city required. Given excess water, the city expanded to the limits of the water supply and beyond. As John Huston said to Jack Nicholson in the movie *Chinatown*, "You see, Mr. Gittes, either you bring the water to LA, or you bring LA to the water."

Los Angeles continued to suck the fresh water from the Owens Valley until the latter was too dry to farm. And still Los Angeles kept growing and searching for water to accommodate that growth. The full story of this quest will appear in chapter 4.

Real Estate

With sunshine to bask in and water to drink, the stage was set for all of Los Angeles's real estate salesmen, hucksters, boosters, and con artists. The immigrant had to be seduced out west and sold a piece of the action, and in turn-of-the-century LA, the art of selling the land was developed to a fine art. The developers had no intention of repeating past mistakes. The boom of the 1880s, for example, had not sustained itself. Between 1884 and 1888, more than a hundred new towns were formed. Sixty-two of these abrupt little burgs became extinct within a few years. By 1889, the boom was dead and gone.

The hucksters of the early twentieth century made plans for the next boom to last. General Harrison Gray Otis organized the Los Angeles Chamber of Commerce, which began a comprehensive selling of the area around the country. As a first act, the chamber proposed that the US purchase Baja California from Mexico and join it with Southern California to form a new state. The chamber also developed a promotional train called "California on Wheels," which rambled through the Southeast and Midwest, carrying photos, displays, and exhibits extolling the Los Angeles way of life. By the turn of the century, Los Angeles was known as "the best advertised city in the country."[3]

The general's successors—Harry Chandler, Henry Huntington, and the well-named Motley Flint—perfected new methods of selling land to the masses. The ephemeral township was replaced by the carefully planned subdivision. In Hollywood, the streets were paved, and a bank and the Hollywood Hotel were built before a single lot was sold. Lots were loaded with building materials to give the impression that home construction was about to begin. Many lots were falsely marked "Sold" before a single one had been purchased.

The Los Angeles real estate boom was the largest, most carefully planned undertaking of its kind in American history. Millions of dollars were gambled that people would continue to move to the area. If the people had turned elsewhere, the promoters would have lost their collective shirts, the construction workers would have been unemployed, and the new subdivisions would have joined their cousins of the 1880s as ghost towns. But the people kept coming. The boom peaked in the '20s, when 1,272,037 people moved to Los Angeles, in what has been described as the "largest internal migration in the history of the American people."[4] The great LA real estate gamble had succeeded.

Transportation

With sunny weather beckoning bedraggled Midwesterners, with water and land awaiting, a way was still needed to get the poor devils out to California. The Southland's first break came with the arrival of the Santa Fe Railroad in 1886. In fact, for many years LA was a rail town; the primary form of local transportation was a municipal electric rail system.

Then, with the early twentieth century, came a momentous twist: the invention of the automobile. Sun-starved Americans, searching for the good life, hopped in their jalopies and headed west. "Like a swarm of invading locusts, migrants crept in all over the roads," wrote Mildred Adams of the Los Angeles-bound hordes. "For wings they had rattletrap automobiles. . . . They camped on the outskirts of town, and their camps became new suburbs." The number of automobiles in Los Angeles jumped even faster than the population, from 160,000 in 1920 to more than 800,000 in 1930.

And the means of arrival began to merge with the way of life in a peculiar auto-erotic dance. Los Angeles quickly abandoned its public transit system and began a love affair with the automobile that has persisted to this day, an affair that has survived the onset of health-threatening pollution and mind-boggling traffic jams. "Our forefathers," editorialized the *Los Angeles Times* in 1926, "in their immortal independence creed set forth the 'pursuit of happiness' as an inalienable right of mankind. And how can one pursue happiness by any swifter and surer means...than by the use of the automobile."[5] This, of course, is a creed that could be applied to all Americans, but nowhere has it reached a loftier state of acceptance than in Los Angeles.

Oil

The poor immigrant, seduced by sunshine and cheap land, watered from a distance, and delivered by automobile, needed a job. The Southland provided this without breaking a sweat. In 1892, a ragamuffin, down-on-his-luck metal prospector named Edward Doheny stumbled into town from the Rocky Mountains. Down to his last few hundred dollars, Doheny moved into a small hotel. One day he saw a wagon loaded with ore pass by. Something about the load piqued his curiosity. "I took a handful and found it was tarry and greasy," Doheny recalled. "I asked the negro driver what it was. 'It is breer,' he replied. That was his crude way of pronouncing '*brea*,' the Spanish word for pitch."[6]

Doheny learned that the greasy ore had come from near West Lake Park. With two buddies, he began to dig a well in the area. The men used techniques that Doheny knew from mining for metals in the Rockies. These had not previously been applied to oil drilling. They worked. After 40 days of digging, the men had a well that produced seven barrels a day of thick oil. Soon, they had 1,500 wells dug and were producing one-third of California's oil.

Doheny, who had started with little more than a pick and shovel, quickly became one of the nation's richest men. He purchased other holdings and drilled other wells, and by 1910 he accounted for the production of 77 million barrels of oil. Ironically, this oil was not of high quality, but was well suited for refinement into gasoline, perfect for fueling all those fancy new cars in the area. By 1927, California had become the nation's leading oil producer. The flood of oil money created jobs and spurred the Southern California boom afresh.

Movies

Just one more historical fluke was needed to give the early-twentieth-century LA boom its distinctive flavor. At the turn of the century, American inventor Thomas Edison controlled several patents on the movie camera. Independent film-makers on the East Coast who chose to ignore these patents were often harassed out of production by Edison's subpoenas and legal maneuvers. These artists tried several distant locales for shooting, hoping to get out of range of

Edison's New York lawyers. Some tried Cuba but found the disease problem too bothersome. Florida was discarded as too warm and humid. San Francisco was used as an early film center, but its climate was considered too rainy. In 1907, a film-maker named William Selig was trying to complete a film version of *The Count of Monte Cristo*. He decided to produce it in Los Angeles, because it was as far as he could get from New York, and because the nearby Mexican border provided a refuge in case of any untoward court orders.

During production, Selig noticed that Los Angeles had a terrific climate, plenty of sunny days for filming, and a startling geographical diversity, sufficient for any number of locales and settings. Word got out quickly. Los Angeles became the center of America's strangest and most distinctive industry. The final ingredient was added to the LA boom: glitter.

Sunshine, water, earth, oil, motion, and glitter: the recipe for a crazy mixed-up historical salad, for a stunning change in the demographics, culture, and politics of an entire state. The pendulum headed south in the early twentieth century, and it has not yet headed back. The LA boom continues. The people in the North try to tackle the problem of their own powerlessness. But by what power can they overcome it? This is the Catch-22 of modern split-state politics in California. The story of regional dissension in modern-day California has been the tale of magic, sleight-of-hand, and desperate action by the North to keep the Southerners from fully realizing and acting upon the power, that by the very strength of inexorable numbers, is theirs to use. For more than half a century, the North has fought a political holding action.

The Federal Plan

The pendulum of history, relentless as it is, has not proved impervious to political efforts to halt or slow its motion. Perhaps the most important of these efforts was the creation in California of the so-called Federal Plan legislature. Reading the results of the 1920 census, many Northerners anticipated that the 1930 census would force a reapportionment that would give Southern California complete control of the state legislature. The Farm Bureau, the Grange, and the San Francisco Chamber of Commerce sponsored an initiative in 1926 to rearrange the legislature. One house, patterned after the US Senate, would be apportioned on a geographical basis. The state senate would include 40 districts roughly drawn on a one-seat-per-county basis. Some very small counties were grouped together; but no county, even Los Angeles, could have more than one senator. Furthermore, no senator could represent more than three counties.

Under this plan, the other house, patterned after the House of Representatives, would be apportioned on a strict population basis. The assembly would be composed of 80 equal districts.

Southern leaders countered with a second plan, under which both houses would be apportioned on the basis of population. Both plans were placed on the 1926 ballot. The southern plan was defeated, 492,923 to 319,456. The Federal Plan, known as Proposition 28, carried by a margin of 437,003 to 368,208. The notion of balancing geography and population, with its long American tradition, proved palatable to the voters.

The Federal Plan drew an immediate and angry response from some southern legislators. They perceived that northern counties could band together to control the senate. Secession spirit flared up in the South one more time. W. F. Beal of Imperial County introduced a bill to form a separate state out of the eight counties south of the Tehachapis. The bill aroused some interest but died in committee in November 1926.

The Federal Plan was instituted just in time for northern interests. The 1930 census did indeed call for a drastic reapportionment of assembly districts as Los Angeles continued to grow by leaps and bounds. As of 1931, 42 of the 80 assembly seats belonged south of the Tehachapis. The Southland established a hammerlock on the California Assembly that it has never relinquished.

To the surprise of some, the Federal Plan actually seemed to work. Population and geography, North and South, city and country all seemed just enough in balance to defuse sectional tension. If the truce was fragile, at least the Northerners reaped a psychological benefit, a peace of mind, a sense that the senate would protect them from a southern juggernaut. This state of affairs led to a sort of *Pax Romana* in the history of California split-state movements, a peace that would last, with one exception, for almost four decades.

The Yreka Rebellion

In the midst of the relatively peaceful Federal Plan interregum of 1926–1965, there arose the most colorful and bizarre of California's split-state movements. The Yreka Rebellion should, by all rights, have faded into the obscurity that has swallowed other rural-based secession efforts. That it did not is a tribute to the mayor of one hick town in Oregon, an enterprising feature-writer from San Francisco, and a large amount of what can only be called strange coincidence.

The story begins with one Gilbert Gable, a proud product of the free enterprise system, a vintage American peregrinator and entrepreneur. Gable drifted

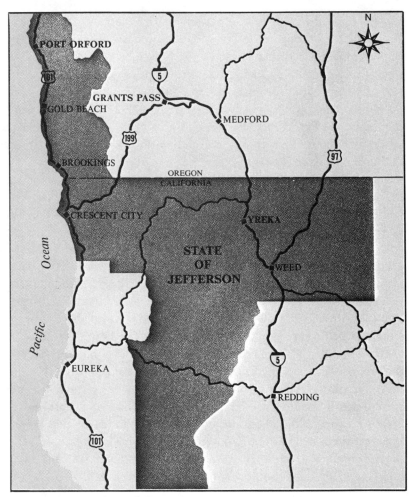

Jefferson, almost the 49th State

through the early twentieth century using his wits, his writing skills, and a fertile imagination. He began as a publicity director for Liberty Loan Drives during World War I. Later, he wrote scripts for radio shows and movies, and spent eleven years as a public relations official for the telephone company in Philadelphia.

In middle age, Gil Gable moved to the tiny town of Port Orford, Oregon. He may have been seeking small town peace and anonymity, a rural retirement from East Coast pressure and the fast-paced business world. When the chance came, however, to run for mayor of Port Orford, this charming, well-groomed, smooth-talking city slicker couldn't resist. Port Orford couldn't resist him, either. He won easily.

As the winter of 1941–42 approached the Klamath River country of far northern California and southern Oregon, Mayor Gable heard some familiar complaints from his constituents. The few roads were in bad shape, people said, and would no doubt wash out again in the heavy winter rains. The people of Port Orford would then go out and try to mend the roads as best they could. Meanwhile, they would suffer through another year of marginal income, of near poverty, thinking all the while about the valuable minerals and timber that lay in the surrounding hills, there for the taking if only some decent roads existed to provide a means of extraction. Vast deposits of copper, tin, chromium, nickel, and manganese lay unmined in southern Oregon. Dense forests of sugar pine, ponderosa pine, and oak lay unharvested.

Mayor Gable had noticed that the price of chromium had been driven from $20 a ton to $48 by the war in Europe. The cost of all raw materials was going up fast. The people of Port Orford were sitting on a gold mine and going broke. That made the damn roads look even worse. And worst of all, Gil Gable and his constituents knew that the folks up in Portland or in the state capital didn't give a hoot about their problem.

Sometime during the fall of 1941, Gil Gable decided that something had to be done. He remembered an interesting tactic that had been tried in 1935 by a group in Crescent City, California. Led by Judge John Childs, this group had threatened to organize several Northern California counties into the separate State of Jefferson to dramatize the area's lack of good roads. Childs's movement didn't get very far, but Gable knew that the idea had possibilities. It just needed the talents of a good public relations man.

In October 1941, Gable and a few friends stormed the Curry County (Oregon) courthouse and demanded legal authority to move the whole county into California! The surprised judge appointed a commission to study the proposal. Gable wrote a letter to California's Governor Olson requesting a meeting to discuss a possible union. The mayor also took care to notify as many local newspapers and radio stations as he could. The word spread quickly throughout Curry County and soon leaked across the border into Northern California.

The problems that Gil Gable had identified in Port Orford were felt just as keenly in the far northern counties of California. Southern Oregon and Northern California shared a common heritage and a bunch of common problems, including the main highway—the tortuous Klamath River Road—a narrow, highly erodible affair with many light-duty bridges. Like Port Orford, many Northern California communities shared the ironic circumstance of a flagging economy amid an embarrassment of natural riches. Like Port Orford, these northern towns felt the palpable indifference of a distant state legislature. Gil Gable's "Good Roads Rebellion" found a home in the northern reaches of the Golden State.

Instead of warmly welcoming Curry County into California, though, Siskiyou, Modoc, and Del Norte counties proved to be much more interested in grabbing the Oregonians and forming a 49th state. The rebellion began to heat up. The capital of the new movement became the rugged mining town of Yreka. This community of some 2400 had its own built-in inferiority complex. The standard joke ran that Yreka did not really exist at all, but was simply a way of misspelling Eureka, the region's largest town. Yreka was a community ready to come out of the closet.

Yrekans thirsted for the publicity that Port Orford had begun to receive. "Hizzoner is one smart cookie," wrote the editor of Yreka's *Siskiyou Daily News* about Mayor Gable. "More people have heard of Curry County in the past month than heard of it before in 40 years." The Yreka Chamber of Commerce voted to investigate the possibility of joining Curry, Modoc, Siskiyou, and Del Norte counties to form a new state called "Mittelwestcoastia." This cacophonous label drove the local paper to run a contest to name the new state. This, in turn, reaped a hideous harvest of Orofino, Bonanza, Del Curiskiou, Siscurdelmo, and other assorted rhetorical nightmares. Blessedly, someone remembered Judge Childs and his 1935 movement. The Yrekans settled on the name of Jefferson for their new state. (By choosing the name of Jefferson, the Yrekans had hearkened back to an era and a presidency when rural America was in its glory. Thomas Jefferson enacted the Louisiana Purchase to double the area of the United States and to open the way to the Pacific Northwest. He had done everything in his power to keep government small and to keep America a land of small farms. "When we get piled upon one another in large cities," Jefferson said, "we shall become corrupt.")

On Thursday, November 27, 1941, the would-be State of Jefferson created a provisional government. Mayor Gable was appointed governor. He was assisted by local State Senator Randolph Collier and by Yreka's mortician, Homer Burton. As governor, Gable stepped up his public relations program to bring the movement statewide attention.

Locals armed with deer rifles stopped autos on the Klamath River highway

Courtesy, the San Francisco Chronicle

and distributed copies of Jefferson's newly written Declaration of Independence. "You are now entering the new State of Jefferson, the 49th State of the Union," read all the tourists and truckers. "Jefferson is now in patriotic rebellion against the States of California and Oregon." Shoppers in Yreka and other communities found "Good Roads" buckets on many cash registers. The idea was to withhold state sales tax pennies from California's government. "No more copper from Jefferson," went the saying, "until Governor Olson drives over these roads and digs it out."

Gable's publicity efforts bore fruit almost immediately. The region's largest newspaper, the *San Francisco Chronicle,* afraid of depressing its readers with too many war-rages-on-in-Europe articles, sent a bright young reporter, Stanton Delaplane, to cover the rebellion. Fortunately for Jefferson, Delaplane was no 5-W's journalist. He was a feature writer at heart and gave the story all the wit, sympathy, and back-roads hyperbole it deserved.

Delaplane's articles sang of the great natural beauty of the Klamath River country. Readers got the impression that, like the homeland of the Seven Dwarves, the northern hills simply bulged with precious gems and minerals. Delaplane told how the ducks on the Creason Farm at Dunsmuir had gold in their gullets from drinking the metal-laden creekwater. He regaled his audience with tales of Buffalo Bill Lang, swashbuckling organizer of Jefferson's state militia. He hinted darkly of the Yrekan tradition of lynchings, including an incident where four people were hung together. "The people here got tired of paying the sheriff so much to feed those four in jail," deadpanned Homer Burton, the mortician.

Delaplane touched on an important ambiguity in Jefferson. These people were kidding, of course. It was surely just a publicity stunt; they were ludicrous rubes, certainly. Yet the Yrekans were also deadly serious; they were tough, hard-working folk with a legitimate gripe and a deep-seated hostility toward Sacramento politicians and particularly toward Southland voters. Delaplane made it clear that while the Yrekans were smiling, they were also armed and potentially dangerous.

BANCROFT LIBRARY

PROCLAMATION OF INDEPENDENCE

You are now entering Jefferson, the 49th State of the Union.

Jefferson is now in patriotic rebellion against the States of California and Oregon.

This State has seceded from California and Oregon this Thursday, November 27, 1941.

Patriotic Jeffersonians intend to secede each Thursday until further notice.

For the next hundred miles as you drive along Highway 99, you are travelling parallel to the greatest copper belt in the Far West, seventy-five miles west of here.

The United States government needs this vital mineral. But gross neglect by California and Oregon deprives us of necessary roads to bring out the copper ore.

If you don't believe this, drive down the Klamath River highway and see for yourself. Take your chains, shovel and dynamite.

Until California and Oregon build a road into the copper country, Jefferson, as a defense-minded State, will be forced to rebel each Thursday and act as a separate State.

(Please carry this proclamation with you and pass them out on your way.)

State of Jefferson Citizens Committee Temporary State Capitol, Yreka

The *Chronicle* writer's colorful reportage helped spread secession fever around the north state. On December 1, the movement peaked. Trinity and Lassen counties joined up, the latter pledging the services of Mt. Lassen, the state's only active volcano. Where actual secession sentiment did not take hold, there grew an amused sympathy. The Associated Farmers of California lent their official support to Jefferson. Surprise Valley in Modoc County astonished everyone by announcing it wanted to become part of Nevada. The *Santa Cruz News* poked fun at the general fluidity of allegiance by suggesting that Santa Cruz secede and become part of Portugal. Secession spirit was in full bloom. Whether or not anyone involved was serious, the Yreka Rebellion had become big news.

The excitement was too much for some, particularly for one of Jefferson's leading citizens. On December 2, Governor Gable up and died, the victim of acute indigestion and extreme nervous tension. But the governor's publicity efforts had given the movement enough momentum to survive his sudden departure.

Fittingly, the venerable Judge Childs of Crescent City was chosen over Yreka Mayor Albert Herzog, a 78-year-old retired Indian fighter, to replace Gable.

Under new leadership, Jefferson geared up to celebrate Gil Gable's greatest legacy, a full-blown media event called Inauguration Day. Hordes of reporters, crews from two national picture magazines, and a newsreel company flocked to Yreka for the occasion. The *Siskiyou Daily News* asked locals to "please wear western clothes if they are available."

Jefferson made the most of its moment in the sun. The girls of the high school drum-and-bugle corps, dressed in brilliant scarlet uniforms, entertained the crowd. A boy in a coonskin cap drifted through with two bear cubs on a leash. Pistol-packing miners carried signs that said: "Our roads are not passable; hardly jackassable" and "If our roads you would travel, bring your own gravel." It was Barnum and Bailey and the Old West rolled into one; the photographers ate it up.

Cannons boomed on the courthouse lawn. A torchlight parade weaved through the streets of Yreka, lighting up the night sky. Governor Childs read a proclamation of secession. Inauguration Day came off perfectly, a well-orchestrated media event, just as Gil Gable would have wanted it. Jefferson's leaders sensed that a rush of publicity would soon be theirs, that Yreka jokes would become a thing of the past, and that new roads were as good as built.

In 1941, though, the time lag between event and mass media presentation was considerably greater than it is today. Newsreels had to be processed, edited, and distributed to theaters. Picture magazines had a long production time. Even newspapers were not as quick as they are today. Before Jeffersonians could experience the sweet publicity almost guaranteed to improve their lot and their road system, cruel fate intervened. Inauguration Day was December 4, 1941. Three days later, the Japanese Navy attacked Pearl Harbor and eclipsed the Yreka Rebellion. America entered World War II, and Jefferson was simply forgotten. The splashy photo essays of happy Jeffersonians marching in bright defiance were buried by war-minded editors. "The newsreels never even got into the theaters," sighed one old-time Yrekan. Even the *San Francisco Chronicle* abandoned Jefferson, as Delaplane rushed back to his desk.

Actually, Stan Delaplane may have been the biggest beneficiary of Gil Gable's "Good Roads Rebellion." The young reporter received the Pulitzer Prize for his lively reports on the rebellion. Some grumbled that he had invented most of the colorful stories and quotes he had reported, but the award stood. Today, more than 40 years later, Stan Delaplane still has a daily column in the *Chronicle*.

For their part, the Jeffersonians were more than patriotic enough to abandon their efforts and prepare for a war effort. Governor Childs announced on December 8 that "the acting officers of the Provisional Territory of Jefferson here and now discontinue any and all activities." The Yreka Rebellion was ended; peace would continue between Northern and Southern California for another quarter of a century.

Jefferson Lives!

The Yreka Rebellion has been ignored by most historians or dismissed as a joke or a publicity stunt. Certainly its leaders never had any clear intention of seceding from California, or Oregon, or anything else. Nor did the movement get much of a rise out of the state legislature or the much-maligned Governor Olson. At the time, the State of Jefferson provided little more than morning reading pleasure for San Franciscans. Some Oregonians showed a bit of pique. "Maybe it's all an advertising stunt," wrote the *Portland Oregonian*, at the height of the rebellion, "but Mayor Gable is carrying his side of the comicability to extremes. . . . [He] has ceased to be funny."

The State of Jefferson had no real immediate impact, it's true, but looking at the history of California by way of secessionist movements sheds new light on many events and historical currents. The footnote can become an earthshaker, the leaky faucet an historical watershed. The Yreka Rebellion reveals a lot about the dynamics of secession politics. First of all, Yreka shows that secession fever is a contagious bug. A full-blown statewide movement is not a necessity; rather, a hot-headed local area is enough. The noise of 1941 began in tiny Port Orford, a hamlet with no telegraph, no railroad line, not even a public library. Within a few days, most of far northern California was involved.

The role the media played in 1941 is equally instructive. Even though only one important journalist covered the story, the media helped legitimize the movement. Diverse and remote areas were able to stay in touch with developments, and media coverage worked to spread the secessionist bacilli throughout the North. All these effects would be magnified greatly today with the larger role of mass media in everyone's affairs. The impact of media coverage on a true secessionist movement could be incalculably large.

The Yreka Rebellion also showed that single-issue politics could provide a working basis for secession. The ignorant, distant, unconcerned population center is the perfect scapegoat; the particular issue lends a banner under which to organize diverse interests. What could happen, one wonders, with an issue more profound than road maintenance—say, water use and water rights? It is not hard to envision a broad secessionist coalition forming under these circumstances.

The Yreka Rebellion was not strictly speaking a North-South dispute, but rather a rural-urban conflict. The Jeffersonians never wanted their boundary as far south as San Francisco, let alone at the Tehachapis. Still, there was widespread, if somewhat bemused, sympathy in the Bay Area for the Yrekans. The villain throughout was the southern-dominated legislature in particular, and the populous Southland by implication. "If this was Los Angeles County," one rebel told Stan Delaplane, "they'd have their roads in no time. They've got the votes but we've got the copper."[7] The issue was mismanagement of a crucial natural resource by a distant, nonrepresentative government. The goal was the creation of a political entity that could be responsive to local needs. Whether or not Yreka was a joke, it represents a classic paradigm of the forces required for a struggle powerful enough to divide the State of California.

Perhaps the most interesting lesson to come out of the brief existence of the State of Jefferson was the utter unpredictability of the back-country people. The strong independent streak in people born and raised in rural areas could be a vital element in a secession movement. A Yreka mechanic working on Delaplane's ailing auto captured the spirit of the Jeffersonians: "You can't tell what folks up here will do," he intoned ominously, "when they got a notion."

The Case of People v. Trees

The relatively peaceful era of 1926–1965 was disturbed only by Yreka and a few minor stirrings. In 1956, the State of Shasta was proposed, a joining of eight far northern counties. This effort lasted but 30 days. It is noteworthy only because it marks the first organized resistance to the exportation of northern water to the south. As the southern developers continued to create new metropolises in the desert, the pressure on northern resources increased. In 1960, voters narrowly approved $1.75 billion in bonds to finance a State Water Project. This created Lake Oroville, the San Luis Reservoir, and the California Aqueduct, a system with the capacity to carry more than 4 million acre-feet of northern water to Los Angeles and giant corporate farms in Kern County. The exportation of Northern California water has remained a key issue in North-South relations ever since.

The mid-century peace between North and South California was not actually broken over water. The immediate issue was reapportionment and the Federal Plan legislature. In the two decades following World War II, the population of California renewed its wild growth. Servicemen, Asians, Mexicans, Okies, and Easterners all flocked to the Golden State. The lion's share, of course, headed south, intoxicated by sunshine, glitter, and opportunity.

In 1940, the population of California was nearly 7 million. It ballooned to 10.5 million by 1950 and reached 15,717,204 in 1960. During this period, nearly 40,000 people moved to the Golden State every month! Opinions varied as to the desirability of this boom. Governor Earl Warren rejoiced, saying in 1948: "We are getting the greatest population bargain of all time." Mayor Fletcher Brown of Los Angeles admitted to being less enthusiastic. "Now don't go quoting me that I want people to stay away," he said, "but I do wish they wouldn't come in such numbers."

In late 1962, California passed New York as the nation's most populous state. Then-Governor Pat Brown was thrilled. To celebrate the event, he declared December 31 of that year as "Population Day." Northern Californians viewed the whole business with a far more jaundiced eye. "The occasion clearly calls for mourning," wailed the *San Francisco Chronicle*, "for a gathering of all the inner resources to withstand this historic buffeting under which the State's once magnificent supply of elbow-room and breathing space has vanished."

As Los Angeles grew and sparkled like a magical mushroom, the strain on North-South relations increased apace. The ever-widening population gap sharpened cultural differences between the areas, heightened the demands on northern water and other resources, and further skewed the balance of political power toward the South. These distinctions and disparities soon coalesced in one of the most critical judicial decisions in California's history.

Since the state senate was apportioned on a roughly geographical basis, the rapid growth of the Southland was not reflected in the number of its senatorial districts. In 1960, for example, the twenty-eighth senatorial district, including Alpine, Inyo, and Mono counties, had a population of 14,294. The thirty-eighth district, Los Angeles County, included 6,380,771 people. Each district had one vote in the senate, so a vote in Alpine County had 450 times the weight of a vote in Los Angeles. Eleven percent of the people of California could elect a majority in the senate. The regional balancing function of the senate had become a glaring inequity in the minds of some.

Despite this inequity, the voters in California had steadfastly supported the system. Four times between 1927 and 1962, initiatives to remove the Federal Plan were defeated. In the 1948 election, Governor Warren made a strong defense of the system. "Many California cities," he said, "are far more important in the life of the State than their population bears to the entire population of the State. . . . Our State has made almost unbelievable progress under our present system. I believe we should keep it." In a 1959 poll, 75 percent of Californians supported the Federal Plan, with Northerners and Southerners contributing in equal measure to the result.

Apparently, Californians from all regions felt that a geographically based senate provided a crucial form of balance, a means of defusing tension between North and South, between city and country. "When 85 to 90 percent of a State's population can be concentrated on less than 2 percent of its land area," warned Richard Carpenter, director of the League of California Cities, "the tyranny of the majority toward minority interests can be as devastating as any exercised by a single dictator."[8]

The Federal Plan gave California what was described as a "rolling consensus" between the assembly and senate. A bill in the assembly that was too disadvantageous toward the North or toward rural areas would be killed in the senate. It was a system that worked by mutual consent, an oafish but effective compromise for a lanky state whose 800-mile north-south expanse is rivaled only by Texas among the contiguous 48 states.

In the mid-1960s, though, time and the US Supreme Court caught up with the Federal Plan legislature. First, the South became more aggressive in its efforts to change the system. In 1960, Frank Bonelli, chair of the Los Angeles County Board of Supervisors, introduced Proposition 15, a plan designed

Renault, *Frontier Magazine*

to give 13 southern counties half of the state's senators. Governor (Pat) Brown threatened that "If Proposition 15 passes, we could have a sectional war in this state."[9] The proposition lost, 3.4 million to 1.9 million. In 1962, however, a resurrected version of the Bonelli plan came close to overturning the federalized legislature, losing narrowly, 2,495,440 to 2,181,758.

It remained, however, for the Supreme Court to effect what California's electorate could not. In 1946, in the case of *Colegrove* v. *Green,* the high Court had ruled five to three that the reapportionment of state legislatures was the business of the states, not the federal government. Justice Black dissented, stating that unfair apportionment was an infringement on the right to vote and that the right to vote is guaranteed by the Constitution. Nearly 20 years later, this dissent would become the majority opinion.

In 1962, the mayor of Nashville asked the Court to call for the reapportionment of the Tennessee General Assembly. Reversing the decision in *Colegrove* v. *Green,* Justice Brennan wrote that the claim that the Tennessee legislature was violating the equal protection clause of the Fourteenth Amendment did constitute "a justiciable constitutional cause of action upon which the appellants are entitled to a trial and a decision." This case, *Baker* v. *Carr*, signaled the end of the line for the Federal Plan.

It was ironic that the Warren Court was the first to rule against the Federal Plan. Earl Warren had been a staunch defender of the plan as governor of California. As a Supreme Court justice, he took a dim view of it. "Legislators," he said, "represent people, not trees." The Court had concluded that, while such plans might serve some arcane intrastate purposes, their overall effect might be to retard social progress, prolong discriminatory practices, and in general to frustrate the will of the people.

In a series of rulings, the Court continued to cite the equal protection clause: "No state shall deny to any person within its jurisdiction the equal protection of the laws." "A citizen, a qualified voter," said the Court in *Reynolds* v. *Sims*, "is no more or less so because he lives in the city or on a farm." A resident of Alpine County could not have a vote worth 450 times the vote of a Los Angeles resident.

California, for its part, tried to argue that its system was simply based on the US model, in which each state has two senators, regardless of population. Why permit such a system on the federal level and deny it to the states? The

© San Francisco Chronicle, 1966. Reprinted by permission

Court's response here is quite instructive and may prove to be significant in any future attempts to reinstitute some facsimile of the Federal Plan. The Court called the state's analogy with the US Senate a fallacious one. The federal system, ruled the Court, is based on a compact among sovereign entities, the states, whose very sovereignty predates the US government. Counties, on the other hand, are not sovereign entities, but are simply districts of convenience created by the states themselves. Because a county is not sovereign, it has no right to representation in a legislative body. The Court permitted bicameral state legislatures to persist, but both houses had to be elected on the basis of population. This is known as the "one man, one vote rule."

In retrospect, it is hard to determine whether the Federal Plan legislature in California was an instrument of social conservatism and disenfranchisement or a crucial element in an interregional truce. Probably it was a little of both. Gladwin Hall, former head of the *New York Times'* Los Angeles bureau, quotes

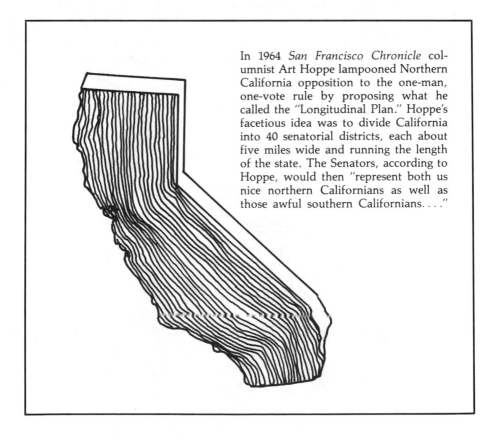

In 1964 *San Francisco Chronicle* columnist Art Hoppe lampooned Northern California opposition to the one-man, one-vote rule by proposing what he called the "Longitudinal Plan." Hoppe's facetious idea was to divide California into 40 senatorial districts, each about five miles wide and running the length of the state. The Senators, according to Hoppe, would then "represent both us nice northern Californians as well as those awful southern Californians. . . ."

"I lost my vote . . . in Sac . . . ra . . . mento . . ."

a veteran senator advising a newcomer to the geographically based body. "Remember, son," the old-timer said, "your job here is to protect the people of California from the acts of the Assembly." Carey McWilliams theorized that during the period between 1928 and 1960, Southland power brokers felt that their lobbyists were actually able to work more effectively in a small body like the senate, where many legislators were responsible only to small, unobtrusive constituencies. "Business interests," wrote McWilliams, "had come to realize that a lack of democracy had its advantages to special interests."[10]

On the other side of the coin, though, many Northern Californians felt that the Federal Plan legislature served a vital function in California. It protected the underpopulated areas and, at least in a purely psychological way, defused much of the regional tension between North and South. Certainly many long-time northern residents mourn the passing of the Federal Plan. "Ever since the Court changed our legislature," says Mill Valley writer Barney Gould, "we've been at the mercy of the Southland." We will return to this question of the effect of California's legislative structure on North-South politics in chapter 6.

The Dolwig Rebellion

No sooner had the "one man, one vote" rulings popped out of the august typewriters of the highest court, than the cry of secession resounded through-out Northern California. *U.S. News & World Report* commented at the time that a secessionist impulse had come to pervade the whole state: "What is be-ing reawakened here, at this time, is the feeling of many that two Californias already exist in fact, if not in law. . . . Even in southern California it is being said that this big and powerful state is too diverse, and has too many feuds smoldering below the surface, to permit unchecked rule by the majority under the Supreme Court formula."[11]

The challenge of concretizing this inchoate secessionist urge was taken up, quite naturally, by an about-to-be-deposed senator. State Senator Richard J. Dolwig, a Republican from San Mateo, introduced legislation to split Cali-fornia into two states. Surprisingly, 25 of the 40 state senators cosponsored Dolwig's legislation. Dolwig's proposal actually involved two bills that to-gether would cleave California at the Tehachapis, pending approval by the assembly, the governor, the state's voters, and the US Congress. The new northern state would have an assembly with 20 representatives from the Bay Area and 20 from the outlying areas. Dolwig argued that six states had already been formed by division: North and South Carolina, North and South Dakota, and Virginia and West Virginia. He appealed to northern legislators that their constituents would suffer under the hegemony of the Southland. He cited the commandeering of the Owens River as an ill omen for the fate of northern resources. "Years ago," Dolwig warned, "we got an example of how Los Angeles exercises its power whenever it can get away with it."

The *Los Angeles Times* took Dolwig's effort seriously enough to respond in a lengthy editorial. The notion of splitting California, said the *Times,* in-volved some very tough questions that Californians needed to face.

What would happen to our master plan for higher education, and with it the University of California, one of the world's truly great institutions of higher learning?

Who would pay for all the highways and freeways built throughout the north in large part with gas tax money from Southern California motorists?

How would the north escape its commitment to those who hold California's bonds, soon to total in the billions of dollars?

What becomes of the vital State Water Plan?

Banks, now operating in both sections of the State, would be forced to make a choice between the north and south. Federal law does not permit interstate banking.

Innumerable businesses presently engaged in intrastate commerce would be transformed into interstate operations, with concomitant imposition of Federal regulations to which they are not now subject.

Every insurance company and other major business or profession operating under State license would become subject to the fees, examinations, supervision, and various regulations of two superintendents, boards, commissions, or what have you, if doing business in both north and south.

The list is virtually endless. And so is the potential for economic chaos resulting from such a whimsical creation of two smaller Californias where one mighty state now exists.[12]

(These very concerns and others like them will be considered in detail in chapter 5.)

Despite this editorial and strong public opposition,* Dolwig's bills were passed by the senate, one by a resounding 27 to 11 margin. For a brief time, northern papers were filled with heady talk of secession. Polls were taken, tentative plans made. But most experienced analysts saw the shallowness of Dolwig's effort. It was not a grass-roots movement. The issue of political organization and apportionment was too abstract, too far removed from people's lives. The emotional elements present in the Yreka Rebellion could only be alluded to by Dolwig.

As expected, Dolwig's bills never got out of committee in the population-

---

*The California Poll of January 1965 indicated that 61 percent of Northern Californians and 76 percent of Southern Californians opposed division of the state.

based assembly. They were killed by the aptly named Interstate Cooperation Committee. James Mills (Democrat, San Diego) and Charles Conrad (Republican, Sherman Oaks) argued forcefully that California needed the prestige of being the most populous state. They added, somewhat speciously, that the North had nothing to fear since no harm had come to it prior to 1926, when the Federal Plan was instituted. This argument did little to reassure Northerners, as the Federal Plan was enacted in the nick of time, before the burgeoning South had begun to assert its political dominance.

Modern Times

With the defeat of Dolwig's plan, secessionist sentiment among legislators faded. Everyone prepared for reapportionment, some even rebuking Dolwig for his efforts. One senator, Alan Short, described the senate's passage of Dolwig's bills as "almost facetious."[13] Dolwig, for his part, managed to fire off one of the decade's worst predictions as he was leaving office: "We won't have the US Senate as now constituted much longer." But Dolwig and others made a more plausible claim, namely, that the issue of water use and exportation would eventually force the secessionist feelings of Northerners back to the surface.

The 1970s saw this prediction come true. Ex-governor Edmund G. (Pat) Brown, stung by his defeat at the hands of Ronald Reagan, penned a book in 1970 that attacked Reagan and the conservative southern power brokers. The book was titled *Reagan and Reality: The Two Californias.* In part, Brown meant to indicate that one California was reality, while the other existed only inside Reagan's brain. A secondary theme of the book, however, was the author's belief that splitting the state was inevitable because "Southern Californians will continue to foist such radical Republicans as Reagan and [Max] Rafferty into authority over the wishes of the 40 percent minority of Northern California voters."

Brown also said, "One generalization is both safe and significant: California is not a cohesive community. Particularly between northern and southern California, the split of the California community is deep, and the gap between those two regions of the state is becoming wider as each year passes . . . .I have reluctantly concluded that California should be divided legally into two states, north and south."[14]

In addition to Brown's manifesto, the 1970s were dotted with secessionist efforts of varying degrees of sincerity.

• In 1970, former state Senator Randolph Collier of Siskiyou County, a veteran of the Yreka Rebellion, urged the splitting of the state down the middle to create the states of East and West California. Collier reasoned that the problems of the state all involved an urban-rural conflict. Since population centers were concentrated along the coast, this form of division would be more appropriate than a north-south split.

• In 1974, a novelist named Robin White led a movement to split the state along a line from Point Arena to Lake Tahoe. This would create a new state called "Mendocino" from the 13 counties north of the thirty-ninth parallel.

This plan was primarily a protest against exploitation of the area's natural resources by the southern part of the state. The movement lasted but a few months.

• In 1978, Assemblyman Barry Keene of Humboldt County introduced a bill to create the state of "Alta California" from California territory north of the Tehachapis. Keene introduced Assembly Bill 2929 to "prevent naked ripoffs of northern resources and trim state government down to a more manageable size."[15]

(The most contemporary efforts to split the state are discussed in detail in chapter 5.)

In the 1980s, the issue of water exportation, as incarnated by the currently defunct Peripheral Canal project, has kept split-the-state feelings very close to the surface. Several different initiatives arrive in the mail. Letters trickle in regularly to the major northern newspapers, calling for the establishment of Alta Libre, or North California, or Ecotopia. The North has been disenfranchised, these correspondents cry out; we must secede. "They will make a desert of the North," writes one man, "to make their own desert bloom." "Five years ago I would have thought the idea ridiculous," says a Stanford University administrator, "but the more we lose our water, our resources, our sense of control over our own lives and destinies, the more it seems that two states may be our only sensible option."

For more than 150 years, California has been threatening to crack apart, to split along a cultural faultline that runs across the state and makes the San Andreas Fault look like small change. The dispute, as it stands now, has two central elements. One is water. The North has it, and the South needs it, or so the most simplistic split-state reasoning goes. The other element is culture. Are Northerners liberal, back-packing nature-lovers, while their Southern counterparts are hopelessly conservative, car-loving over-consumers? Is the Southland the real repository of California culture, while a bunch of rubes and hippies and under-achievers inhabit the North? Have time, place, and historical accident created two diverse tribes North and South, such that a divided state is inevitable? Before we can look at the pressing issue of water development, the idea that Californians inhabit two different cultures bears closer examination.

3
A Tale of Two Cultures

In 1943, *Saturday Review*, bastion of snobbish Eastern culture, broke new ground with an entire issue on California and its culture. The crucial element of California culture, the editors explained in an introductory essay, is the state's peculiar bifurcation. "There are of course the two Californias of north and south," the article affirmed, "as regionally characterized by backgrounds, sentiments, and practices." The essay went on to describe the North as a land of ancient redwoods and of people "infused with Anglo-vitamin vigor."

The Los Angeles area was referred to as the "unctuously captioned Sunny Southland," land of "the leisure-seeking and home-buying tourist, of the glamorous Hollywood firmament."[2]

Nearly every serious observer of California culture has noticed the existence of two separate subcultures. In Joel Garreau's *Nine Nations of North America*, the author opines that "San Francisco and Los Angeles are not just two cities. They represent two value structures. Indeed they are the capital of two different nations."[3]

The notion of irreconcilable cultural differences between North and South has

> "An economist told me, 'In Southern California the tendency is to buy cheap fashionable clothes, cheap houses, expensive cars—and drive off to Las Vegas. In Northern California the tendency is to buy cheap, durable clothes, build a snug and somewhat expensive home, and drive a medium-priced car for five years.'"
> —Eugene Burdick

itself become part of California culture. This is especially evident in the humor, as the two regions delight in making jokes at each other's expense. In early 1982, a conference on the Peripheral Canal was held at UCLA. The audience was primarily composed of Southlanders, but many of the speakers were from the North. One of the speakers was droning on in praise of the remarkable conservation efforts made by Marin County residents during the serious three-year drought in the 1970s. His clear implication was that if Southlanders could do as well, all our water worries would be over. One sharply dressed woman jabbed her similarly chic friend and remarked, "Sure they conserved. They all had their BMWs washed in Oakland!"

More commonly, though, the jokes flow in the other direction. San Francisco morning disc jockeys Mike Cleary and Frank Dill of KNBR (the "Good Times" station) have been trying for months to collect funds for their pet project, the Peripheral Garden Hose: 400-plus miles of plastic hose from their radio station's washroom sink to Los Angeles City Hall to bring Southlanders "all the Northern water they so richly deserve."

"I don't think I ever before realized the distinct difference
between northern and southern California."

Drawing by W. Miller; ©1967 The New Yorker Magazine, Inc.

As a variation on the how-many-Marin-County-residents-does-it-take-to-screw-in-a-light-bulb-joke (Ten: one to do it and nine to share the experience), Santa Cruz author James Houston offers up the following:

"How many southern Californians does it take to make a cup of instant coffee?"

"I give up."

"Two. One to add the protein-enriched, simulated dairy supplement made of soybean concentrate, acidophilus culture and brewer's yeast. And one to steal the water."[4]

San Francisco Chronicle columnist Herb Caen has practically made a career out of poking fun at Southern Californians. "A true San Franciscan," he writes, "gets jet lag flying to Los Angeles." He calls the state's largest city "Louse Angeles" or even "Lozangeles" as though it were a troublesome troche stuck in his Northern California throat.

Caen is only half-kidding, though. He really believes in San Francisco and has little doubt of its cultural superiority. Here is a typical diatribe, written while Caen was in Los Angeles to accept an award.

"Is there still a rivalry between Los Angeles and San Francisco? This is the subject of an exhaustive—and you can bet exhausting—article to appear in the *Los Angeles Times*, a newspaper that also wonders whether it is great or simply big. In trying to answer the reporter's questions, I suggested that the competition is a thing of the past. If size is the criterion, Los Angeles won the war long ago. The big power and money are down here. Unless you are addicted to the cooling fogs, fresh off the Pacific, I suppose the L.A. weather is better, too. The crucial difference for me is that San Francisco has a heart, a center and a heritage of excitement that seems as vital as ever. Los Angeles crawls, bumper to bumper, while San Francisco leaps about its hills like a goat. Los Angeles lacks visual excitement, except for the endless carpet of lights at night. It has beautiful lawns, not to mention Forest Lawn, but so does Sacramento. In San Francisco, nostalgia is a slightly deadly way of life, but valid. Down here, people puddle up over the Carthay Circle Theater and Cecil B. DeMille directing a cast of thousands. The golden age was Louis B. Mayer, a man who had less charm than a cold cheeseburger."[5]

Those who might expect Caen to be a prophet of the split-state movement, however, will be disappointed. "Personally, I think the two-state idea is ridiculous," he complains. "It's wonderful to have two cities so far apart so close together—Californians all." And while Caen is proud of San Francisco and some outlying hot spots, he is usually scathing in his indictment of northern backwater towns, especially Chico, famous for carrying Velveeta in the gourmet section of the supermarket.

Willingly or unwillingly, Herb Caen is the spokesman for a pervasive form of Northern California folklore. This form sees the South as a wasteland of fast food restaurants, condos, and shopping malls. As Phyllis Theroux writes, "Within the state there still exists a natural antagonism between the northern and southern halves, and until my cousins and I were grown, nobody lived in Los Angeles, which somebody in my family described as roughly comparable to floating down a sewer in a glass-bottom boat."[6]

It must be admitted that Southlanders don't spend as much time or energy on this whole business as Northerners do. The Herb Caen of the South is Jack Smith of the *Los Angeles Times,* who claims he's "just a man who writes about his dog and his wife." Caen himself complains that Smith "writes and speaks so fondly of San Francisco that sometimes I think he is being condescending." Southlanders seem to have a quiet sense of superiority that may come from their political dominance. Some of the politicians may even enjoy the envious barbs of northern nabobs. A 1966 magazine article on then-Los Angeles mayor Sam Yorty described this attitude: "Like Los Angeles itself, which has long put up with the patronizing attitude of northern neighbor San Francisco, he seems to take pleasure in playing the underdog even when he knows he is top dog."[7]

To the extent that Southerners partake of this cultural comparison, it is usually over the issue of professionalism. That is the ultimate compliment in the South, to be called a pro. Northern work is likely to be amateurish, they say. If you want a job done right, better do it in the South. By and large, though, Southlanders answer northern jabs by silently enjoying their own benevolent climate; living well, they might say, is the best revenge.

How serious is this word play? How well-founded is it, for that matter? Are the political differences between North and South explicable on a cultural basis? Or rather are political problems exacerbated by this sense of cultural rift? Are we really divided by backgrounds, sentiments, and practices? The best way to find out is to take a close look at the Two California cultures. We have chosen to examine these areas: architecture, art, the car, film, journalism, literature, politics, sociology, sports, and tourism.

Comparing cultures is a speculative business, at best. In this case, comparing micro-cultures within a subculture, it amounts to a bit of a guessing game, like "Let's play Split the State!" or "California Feud!" So come along, if you will, with a bit of carnival spirit, and we will toss around some generalities and compile a sort of cultural encyclopedia of the Two Californias, all with the lofty purpose of discovering whether cultural differences between North and South are keeping California divided.

Architecture

Northern California building style tends toward the eclectic, a mix of Victorian, colonial, mission, and various European styles. It aims, generally, to charm the eye. On the surface, Northern California architecture has an edge over that of Southern California, where tract homes, high-rise office buildings, and condominiums dominate the landscape. Surprisingly, though, in a history of architecture text, the South might receive more attention.

In the early twentieth century, many talented architects gravitated to Los Angeles. San Francisco was viewed by these artists as too restrictive, overly committed to tradition and nostalgia, locked into a Victorian mold. R. M. Schindler, Richard Neutra, the Greenes, Frank Lloyd Wright, and others

William Oliver house, Los Angeles, 1933. Designed by R. M. Schindler, the house had a wood stud frame which was modified to permit a wider use of large glass areas. The living room was furnished with Schindler's unit furniture (see next page), which could be assembled in a variety of combinations. Schindler felt that furniture should merge with the house, leaving the room free to express its form. —from *Five California Architects*

made Los Angeles the hottest spot in the country for architectural ideas in the '20s, '30s, and '40s.

These innovators used materials like steel, glass, wood, and concrete in fresh ways. They scoffed at tradition. Each building was a new problem to be solved. Schindler wrote to Neutra about Frank Lloyd Wright's work: "It has no tradition to overcome or prejudice to fight. His work grows quietly out of itself. He is the master of each material, and the modern machine is at the base of his form-giving."[7a]

Of course, these architects did not work only in Los Angeles. The Greenes worked all over the state, Wright all over the nation. Still, in those heady mid-century decades, it was Los Angeles that stood for innovation and style in architecture.

Ironically, Neutra and the others may have laid the foundation for the desecration of their own work. Their fascination with machines and modernity led them to develop techniques and technology for repetition of design. Neutra himself railed against the classicists' "bias against repetition." These great artists never anticipated the kind of explosive population growth that would bring about an architecture based almost solely on repetition.

M. Louise Stanley, "The Polite Society"

Art

Just as urban sprawl and mass culture were diluting the vitality of Los Angeles architecture in the late 1950s, an innovative artistic community began to spring up there. These LA artists joined the Pop movement of Andy Warhol and Roy Lichtenstein in challenging the Abstract Expressionism that had dominated the New York art world for more than a decade. Los Angeles visual arts came to be characterized by a cool, clean geometry, a style similar to that of the above-mentioned architects.

No one epitomizes Los Angeles art as well as Edward Ruscha. His paintings, drawings, prints, and books of photographs are an encyclopedia of Southern California life. He'll use all sorts of materials—gunpowder, spinach, cherry juice. His most well-known print, "Hollywood," shows a range of hills at dusk with the famous Hollywood sign ambiguously stretched across the sky. Is this a dawn scene, a hopeful omen of opportunity, or a sunset, the fade-to-black of a dying civilization? LA artists are rarely shy; they revel in the very infirm-ities of their beleaguered town. "I've been influenced by everything about

Edward Ruscha, "Standard Station, Amarillo, Texas," oil on canvas, courtesy Hood Museum of Art, Dartmouth College, Hanover, NH

LA," says Ruscha, "all the street iconography, the decadence of life, even the smog."[8]

It appears that a California art style has developed, but within that style there is a distinct difference between northern and Southland art. One of the most important differences has its roots in economic factors. Los Angeles is the second-largest painting and sculpture market in the United States. San Francisco is far behind. "There's not much of an art market in Northern California," claims Oakland artist M. Louise Stanley. "And that means there's not as much pressure to conform. That, in turn, enables artists to be more honest. This is an atmosphere very conducive to making art."[9]

Interestingly, a similar distinction has developed between northern and southern theater. Los Angeles productions are regularly scouted by representatives of the film and television industries. The performers, writers, and directors often have lucrative movie or TV contracts on the line when they perform. This causes them to favor traditional vehicles. "We see fewer risks in theater in LA," says author Barbara Isenberg, "because actors simply look better in traditional plays."[10] By contrast, San Francisco performers are usually free of such pressures. Experimental theater, then, flourishes in the North.

A 1982 San Francisco gallery show offered a chance to view a representative northern artist and a Southlander side by side. Sculptor Stephen de Staebler (from the North) revealed bronze forms cast from clay originals. This form gave the bronze a striking organic quality, like fossil prints in metal, shadows of living things. One reviewer saw in de Staebler's work a "close collaboration between art and nature." De Staebler's partner in the show, Southland artist Ron Davis, works with clean, cool geometric forms and painted surfaces. The pieces are slick and speak of high technology. (Davis's pieces were not well received by the northern reviewer.)[11]

Automobiles

It could be argued that the automobile dominates all of California, both North and South, and there would be a good deal of truth to it. The extent, however, to which the auto permeates Southland culture is remarkable. Eugene Burdick remarked, "The South is the place in the world so geared to the automobile that if it were eliminated the whole region would collapse."[12] To understand Southern California culture, you must understand life based on the car.

In the fast lane of the freeway, the car is home and refuge, defense mechanism and weapon, personal extension and life force. The best expression of this distinctly Southern California relationship to the auto and the road is the character Maria Wyeth, spacey heroine of the Joan Didion novel, *Play It As It Lays.*

> It was essential (to pause was to throw herself into unspeakable peril) that she be on the freeway by ten o'clock. . . . She drove it as a riverman runs a river, every day more attuned to its currents, its deceptions, and just as a riverman feels the pull of the rapids in the lull between sleeping and waking, so Maria lay at night in the still of Beverly Hills and saw the great signs soar overhead at seventy miles an hour *Normandie ¼ Vermont ¾ Harbor Fwy 1.* Again and again she returned to an intricate

Edward Ruscha, "Hollywood." 1968. Screenprint, 17½ × 44½"

stretch just south of the interchange where successful passage from the Hollywood onto the Harbor required a diagonal move across four lanes of traffic. On the afternoon she did it without once braking or losing the beat on the radio she was exhilarated, and that night she slept dreamlessly.[13]

While the people of the Southland may have a more intimate cultural relationship with their cars, statistics show that Northerners do just as much driving, perhaps even more. Highway miles (1980 figures) in the South amount to 6063 miles per person per year. In the North the figure was 6719 miles per person. Southlanders own 640 cars per 1000 people; Northerners 672 per 1000. Both regions suffered about 800 injury accidents per 100,000 people in 1981. The likelihood of a fatal car accident, however, is surprisingly higher in the North. The South experienced 16.5 fatal accidents per 100,000 people in 1981; the North had 22.8 per 100,000.[14]

The real problem with cars in California, of course, is smog. The Bay Area, including San Jose, has its share of air quality problems, but the situation is most acute in Southern California. A recent study indicates that LA smog has a higher concentration of carcinogenic substances than is allowed under federal *water* quality standards! The latest horrid specter of acid fog along the Orange County coast is no more comforting.

Northern distrust of Southland water use may be exacerbated by the way Los Angeles has handled its smog problem. LA is faced with a regularly poisonous purple atmosphere (dispersed only by an occasional, fortuitous burst of wind off the eastern desert), dying emphysema patients, and cancelled physical education classes in school, and still not enough is done to counteract the problem. Some Northerners must be thinking, with a certain amount of justification: "Those people are crazy down there; they'll take our water until they've created a desert, and they'll never even notice that anything is wrong."

There are signs, however, that the South may finally be coming to grips with this serious problem. A recent Field poll shows that all Californians are driving less. Answering that they drove less than the previous year were 66 percent of those in Los Angeles and Orange County; 66 percent of the rest of Southern California; 63 percent of the San Francisco Bay Area; and 66 percent of the rest of Northern California.

More significantly, in the summer of 1982, the Los Angeles Air Quality Board announced a plan for meeting federal standards by 1987. Some of the plan's elements had a curiously Northern Californian, Ecotopian ring: electric cars, ride-sharing, encouraging living closer to work, riding bicycles, using computers and telecommunications for business conferences, and commuter railways. Other methods involved towing airliners into take-off position, synchronizing traffic signals, and restricting industrial pollution (a critical part of LA smog).

The *San Francisco Chronicle* implied in an editorial that a concerted Southern California effort to clean up the air would go a long way toward easing North-South tensions. "Traditional Northern California scorn of the Great Southland has changed in recent decades to acknowledgment of growth and substance. And, of course, to fear ever since one-man one-vote decisions abruptly changed the upstart into the prime political decisionmaker." After that brief jab, the Chron heaped heavy praise on the new smog plan as a superb exercise in self-discipline, something the paper implied was needed in large doses down south. The editorial concluded that "the area will deserve universal admiration if self-discipline succeeds."

Film

The development of Hollywood as the world's film capital certainly helped fuel the extraordinary growth of Southern California. This growth, in turn, has contributed to divisive feelings in California. Does the impact of the movie industry on split-state sentiment run deeper than this?

Without doubt, movies are big business in California. The major studios— Warner Brothers, Columbia, Paramount, Twentieth Century Fox, and MGM/ United Artists—control a good part of the world entertainment industry. Corporate profits for these companies in 1979 amounted to nearly $500 million.

The North has long coveted a share of this colorful industry. (As, of course, have other areas of the country; New York and Chicago have recently made inroads into Hollywood's dominance.) In fact, Northern California's own film industry has a long and prestigious history. The first moving pictures ever were taken in Northern California. In 1877, railroad baron Leland Stanford made a friendly wager on whether all four of a horse's hooves were off the ground in mid-stride. To prove his point, Stanford brought Eadweard Muybridge to his Palo Alto ranch to try an experimental technique for taking moving pictures. The results were the first movies and proof that a horse is indeed momentarily airborne while running.

At the turn of the century, when film-makers were moving west to avoid Edison's lawyers, San Francisco and Oakland had brief reigns as film capitals. Dozens of features were made in the North, including several by Charlie

Chaplin. Eventually, though, film-makers found the Bay Area climate too rainy for year-round work. The industry moved south.

Recently, Northern California's film industry has experienced a renaissance. One of its key figures is Francis Coppola, director of the *Godfather* films, *Apocalypse Now,* and *One from the Heart.* When Coppola was riding high financially, he made a bold move to create a major studio in San Francisco. He also bought super-hip *City* magazine and radio station KMPX, and for a time seemed to be leading a one-man cultural renaissance in the North. The director has since fallen on some difficult times, and his future role in establishing the San Francisco Bay Area as a film capital is unclear.

Other leading Northern California film-makers include Coppola's friend Carroll Ballard (director of the *Black Stallion*), animator John Korty, producer Sol Zaentz of Fantasy Films (*Lord of the Rings*), Les Blank (*Burden of Dreams),* and Jon Else *(The Day After Trinity).* The most important northern film-maker is probably George Lucas, who is responsible for many of the highest-grossing pictures of all time, including *Star Wars, The Empire Strikes Back, American Graffiti,* and *Raiders of the Lost Ark.* Lucas has constructed a multimillion-dollar facility in northern Marin County, including a sound stage, Industrial Light and Magic (his special-effects organization), and a cinematic computer center called Sprockets. With the influence of Lucas and others, the North may be on its way to rivalling Hollywood as a center of cinematic arts and industry.

Journalism

Over the years, one of the hottest split-state rivalries in California has taken place between the two biggest newspapers. Some even mark the border between North and South at the place where the *Los Angeles Times* begins to replace the *San Francisco Chronicle* in the state's driveways.

The *Chronicle* is a feature-oriented paper, not known for its hard news focus or its investigative journalism. It relies on witty columnists such as Herb Caen, Art Hoppe, and even the venerable Stanton Delaplane, who covered the Yreka Rebellion more than 40 years ago and is still meeting his deadlines. The propensity for running trendy features in the *Chronicle* was memorialized in the film *All the President's Men,* when Jason Robards, as Bed Bradlee of the *Washington Post,* claimed that the Chron would run yesterday's weather forecasts.

The Chron has also won a perverse sort of renown for its spectacular typos, solecisms, and editorial gaffes. The paper is a regular contributor to column fillers in *New Yorker* magazine. These fillers are generally scathing recountings of journalistic bloopers from around the country.

UH-HUH DEPARTMENT

(Charles McCabe in the *San Francisco Chronicle*)

"Apart from talent there was one salient difference
between Montaigne and I."[15]

The *Chronicle* did make one interesting departure from its breezy format in the spring of 1982. It ran an extensive, detailed investigative series on a major political issue: the Peripheral Canal. The Chron did its best to ensure that northern voters were armed with every imaginable argument against the canal. Some environmental activists, like David Nesmith, who led Friends of the Earth's work on the canal, feel that the paper went too far in whipping up anti-Southland bias on the issue. The results, however, are inarguable: 92 percent of all Northerners voted against the canal.

The *Los Angeles Times* is a very different sort of newspaper. It is relatively dense, known for in-depth investigative reporting. Its feature writers are pretty dreary; even its film reviewers are average. Political issues are its bread and butter, but the *Times* shies away from one issue, the very one that the *Chronicle* is covering most thoroughly: water politics. One northern writer, a free-lancer whose pieces are frequently carried in the *Times*, recently submitted a piece on water development. It was summarily rejected. "They're just not willing to open up on that issue," complains the writer.

The *Times* supported the canal but never tried to counter the extensive information campaign of the *Chronicle*. The day after the June election, though, the *Times* ran a cartoon that depicted a burly Northerner urinating across the border on his southern neighbors. This struck many observers as a surprisingly cheap shot.

The *Times* may be caught in a bind on water issues. Its parent corporation is controlled by the Chandler family, which, in turn, controls the Tejon Ranch Company, one of the giants of Kern County. Tejon Ranch, at some 250,000 acres, is the largest contiguous farm holding in the state. Since the days of William Mulholland, the Chandler family has been deeply involved in the water exportation business. Because of these connections, one of California's best newspapers has occasionally found itself tongue-tied on the state's most pressing issue.

Winds of change, though, are blowing through the *Times'* offices. The first signal occurred in the *Times'* editorial endorsing the construction of the Peripheral Canal. The paper felt it necessary to deny any self-serving in its support of the project. The editorial revealed the connection with Tejon Ranch, noting that "The holdings in Tejon represent a minor part of Times Mirror assets," and asserting that "The Times arrives at its views on the canal, as on any other issue, utterly without regard to the views of the Tejon Ranch Company."[16]

Then, in November 1982, the *Times* was struck by the reality of its own independence. Someone in the editorial offices realized the paper *could* oppose water development. An initiative called Proposition 13 was on the upcoming ballot, a measure also known as the Water Resources Conservation and Efficiency Act. It proposed to solve many of the state's water problems by altering state subsidization of water pricing and by mandating special measures to protect endangered groundwater basins. (See chapter 6 for a complete discussion of these problems.) As expected, powerful business interests in the state lined up against the measure. The *Times* was expected to follow suit, but the paper surprised northern observers by supporting Proposition 13 in an editorial, which concluded:

Logo courtesy of New York Magazine Company, Inc.

Courtesy San Francisco Magazine

"For most of this century, there was enough money to develop water programs and enough water to sustain them. There was no need for controls. Now the state has entered a new phase of development—one that may well last as long as the first phase when today's systems were built. Proposition 13 is a contract with the future. California voters should not hesitate to sign it."

Prop. 13 was soundly defeated, but the *Times* endorsement stands as an interesting signal, much like a show of detente from Russia or China, that North and South California may be drawing closer together on water use issues.

The one state-wide magazine, incidentally, is less provincial than the dailies. *New West* (now *California* magazine) has done a consistently good job of presenting both sides of Two California issues. A 1977 article called "California Split," by Jeannie Kasindorf and others, is quite interesting. The magazine's coverage of the water controversies leading up to the Peripheral Canal vote is a classic example of balanced but lively journalism.

Literature

Both Californias can be justifiably proud of their literary traditions. The North can claim such literary lions as Steinbeck, Jack London, and Frank

Courtesy Los Angeles Magazine

Courtesy San Francisco Magazine

Norris; quirky talents like Gertude Stein, Ambrose Bierce, Dashiell Hammett, and Bret Harte; poets like Robinson Jeffers, Gary Snyder, and Richard Brautigan; the beats, Kerouac, Ginsberg, and Ferlinghetti; the Stanford set, including Wallace Stegner, Scott Momaday, and Ken Kesey; and other fine contemporary talents like Sam Shepard, Cyra McFadden, Ishmael Reed, and Martin Cruz Smith.

The South can counter this lineup with a list of Zane Grey, Aldous Huxley, Christopher Isherwood, Will Durant, Upton Sinclair, Thomas Mann, James Cain, Nathaniel West, Henry Miller, and Ray Bradbury. Looking at these lists, the only really solid generalization one can make is that literary talents are drawn to the South by sunshine or the movie business and end up sticking around.

Viewed from the perspective of California as the subject of literature, the Southland has played a much greater role than the North. When out-of-state writers try to depict California as a whole, they usually end up with a picture of Southern California. Even with California writers, the Southland seems to make a better dramatic setting. Los Angeles serves as the incarnation of the California dream in all its splendor and in all its wretched excess. And Southern Californians seem to make great characters, too, from the jealous husbands in Raymond Chandler novels, driven to violence by the arid Santa Ana winds, to Maria Wyeth, Joan Didion's child of the freeways.

Northern Californians should probably consider the role of literary sacrificial lamb the Southland has played before they consider splitting the state. Could San Francisco really bear the scrutiny of Eastern writers if LA weren't down there outdoing SF's every kooky turn?

As writer James Houston puts it, "Observers nationwide who scoff at and ridicule what appears to be going on in California are usually scoffing at and ridiculing LA. . . . For the rest of California it is out in front taking the body blows, a kind of cultural punching bag."[17]

Politics

When he wrote the book *Reagan and Reality* in 1970, ex-governor Pat Brown concluded that political ideology more than any other factor separated the Two Californias: "The great difference between northern and southern California is not just in physical characteristics and circumstances, creating different problems and needs for the two areas. The deepest difference is in the political attitudes of northern Californians and southern Californians. That political difference is as pronounced as between Minnesota and Mississippi." [18]

The split-state folklore holds that the North is liberal, the South conservative. This was certainly true at one time. In the presidential election of 1960, for example, the North went for Kennedy 51 to 48 percent, while the South voted for Nixon by the same margin; Nixon carried this important state by 35,000 votes and came within an eyelash of winning the presidency. In 1964, conservative US Senate candidate George Murphy rolled up a 350,000-vote plurality in the South to more than offset the tally in the North, where he lost by 150,000 votes.

Figures like those have become ingrained in folklore and lead some to expect North California to become a liberal haven if it were freed from the yoke of the conservative South. The North could ban nuclear power plants, enforce anti-pollution laws strictly, and legalize marijuana, according to this reasoning. Actually, the Two Californias seem to be converging politically.

"The conservativism of the South has narrowed," says pollster Mervin Field. A glance at current voter registration patterns is illuminating. In 1982, roughly 54 percent of Northern California's registered voters were Democrats, 31 percent were Republicans, and 15 percent registered as Independent or "other." In Southern California (based on the ten-county split we have used for most statistical purposes), 51 percent registered as Democrats, 37 percent as Republicans, and 12 percent as Independent or "other."

Apparently, the old stereotypes of California politics are no longer useful. The state defies analysis anyway, on liberal-conservative lines. Californians elected Ronald Reagan and Jerry ("Governor Moonbeam") Brown as consecutive governors. In 1976, the voters were so likely to call for strict nuclear safeguards that the legislature preemptively enacted less stringent measures only days before the election; the initiative lost, but the idea won. In 1978, voters supported the Jarvis-Gann initiative. In 1982, the voters defeated both the Peripheral Canal and the Water Resources Efficiency and Conservation Act, elected Pete Wilson and George Deukmejian, and endorsed the Nuclear Freeze. Who can figure it?

Perhaps the one conclusive political distinction that can be made between North and South is that the influence of environmentalism is considerably stronger in the North. Northern California is, after all, the birthplace of the modern-day environmentalist movement. San Francisco lists some 25 environmental and conservation groups in the yellow pages of the phone book, including the national headquarters for two of the most powerful of such organizations—the Sierra Club and Friends of the Earth (Los Angeles lists twelve).

The key to the overwhelming defeat of the Peripheral Canal in the North was the activism of a multitude of conservation groups and their volunteers, acting to inform and mobilize voters.

The North is literally crawling with nature lovers—people who like to save rivers and forests, or whales, seals, otters, and just about anything else that breathes. They are unabashed in their opposition to James Watt, Anne Gorsuch, and the other homewreckers of the Reagan administration. Through fat and lean economic times, through the subtle shifts of the pendulum of power in California, the Northern California environmental movement has grown in scope and influence for several decades. Until this movement either waxes in the Southland or wanes in the North, environmental issues will continue to divide the state.

Sociology

In 1977, *New West* magazine looked at the habits of the Two Californias. Northerners, the article found, were more likely to commit suicide, contract gonorrhea, or die in a car accident. Southerners were found more likely to be murdered, raped, robbed, or assaulted, contract syphilis, or have their cars stolen.

Recent surveys show that more Northerners think of California as "one of the best places to live" (72 percent to 68). More Northerners are native Californians (41 percent to 36). Southerners are more inclined to want to live in another state (31 percent to 26; Oregon, Hawaii, Colorado, and Arizona are the leading alternatives).

Southerners seem considerably more confident about their financial futures, according to the California Opinion Index:

Q. Next year I expect to be financially . . .

	Better off	The same	Worse off	Who knows?
San Francisco	30%	33%	22%	4%
Other Northern Californians	25	46	23	6
Los Angeles/Orange County	40	41	15	4
Other Southern Californians	43	35	19	3

After the Peripheral Canal vote, the air was thick with sociological statements explaining the North-South split. One of the most concise came from Stewart Brand, former Merry Prankster, engineer of the *Whole Earth Catalog,* and now editor of the *Co-Evolution Quarterly* (a Northern California magazine if ever there was one—it doesn't even carry advertising!). "My own version of this issue," said Brand in Sausalito, "is that if you want to be successful you go to Los Angeles. If you want to be happy you go here."[19]

Sports

The people of both Californias, North and South, share an almost fanatical devotion to physical activity. Surfers, sailors, volleyball players, runners, and tennis players occur in large numbers in each half of the state. A look at rivalries in organized sports between the two regions, however, may shed some light on cultural differences.

At the high school and college levels, most sports rivalries are local. Cal and Stanford are bitter rivals, as are USC and UCLA, but the intensity of feeling does not survive a trip across interregional lines. Professional sports are a different matter. They are a socially accepted vehicle for venting just barely sublimated antagonisms. Like gladiators in the Roman Colosseum, California's professional sports teams are often unwitting bearers of the banner of regional pride. These teams engage in some of the most ferocious rivalries in all sport, such as the baseball wars between the Giants and Dodgers in the National League and the Oakland A's and California Angels in the American, or the rivalry of the 49ers and Rams in football. When these teams play, the games become little morality plays on the split-state theme.

This element surfaced clearly during the 1982 baseball season, when the Giants and Dodgers battled for the pennant down to the last days of the season. The last three games were played in San Francisco, attended by hordes of howling fans. The Dodgers beat the Giants twice; then, in the final game, the Giants beat the Dodgers, giving the Atlanta Braves the division championship. The Bay Area fans celebrated as if their team had won the World Series. One fan summed up the Northern California attitude. "It's a way to kill LA," said Ron Bistolfo, a musician from San Jose. "I've always hated everything from LA, including their need for water. I hate the Rams, Lakers, and Dodgers."[20]

Over the years, Southern California teams have been more successful, winning 33 major sports championships, to 19 for northern teams. Ironically, the most successful northern franchise has itself become a symbol of split-state antagonism: the Oakland, er, Los Angeles Raiders. Owner Al Davis used some complicated legal maneuvers and a truckload of chutzpah to move his team to Los Angeles and leave a few hundred thousand loyal Bay Area fans out in the cold. Interestingly, some of the anger of these abandoned fans has been directed at the City of Los Angeles for stealing the team, much as many of those same people probably believe the city guilty of stealing water. During the 1982 season, the Raiders lived and trained in the North and flew south to play their games. For this performance, the authors of this book would like to name the Raiders as the official team of the split-state movement. Perhaps Al Davis would consider renaming the team: the Two Californias Raiders.

Tourism

Though the tourist dollar is important to both northern and southern economies, the two regions approach the tourist industry quite differently. In the South, the tourist is deluged with *things to do.* Theme parks abound. Disneyland is the most famous, but there are many rivals, such as Knotts Berry Farm, Magic Mountain, Lion Country Safari, and Marineland. There are studio tours (in fact, the Universal tour is much like a theme park) and tours of the stars' homes. Southern tourism is predicated on artifice and diversion.

In the North, tourism is, well, more existential. The emphasis is on *places to be;* what you do once you're there is left up to the individual. In San Francisco, the big attractions are places, like Fisherman's Wharf, Chinatown, and Golden Gate Park. These are real places where life goes on, not theme parks. Of course, there is plenty to do in San Francisco, but the tourist's role in deciding what to do is a more active one.

Outside San Francisco, northern tourism is equally existential: the Mendocino redwoods, Yosemite Valley, the wine country. Perhaps this different brand of tourism connects with a Northern view of humankind's relationship to nature. The idea is to live as harmoniously as possible within this bountiful natural world; tampering by technology should be minimized. In the South, the dominant worldview is that the tampering is what makes life itself possible. The theme park is just a ritualized version of the State Water Project, a celebration of technology that charges admission.

*   *   *

Culture in both North and South California springs from a common source of ideas. Lurking behind northern provincialism and "era of limits"/"Spaceship Earth" thinking is the powerful California sense that all things are possible. Buried somewhere just beneath the concrete of Southland freeways is a strong bond with nature that is not broken but simply deferred. The Southern California surfer may be searching for thrills and adrenaline rushes in part, but another aspect of the attraction is the peace, natural beauty, and closeness to nature found out beyond the breakers. There are Two California cultures, yes, but they feed off each other, enrich each other, push each other on. It is hard to imagine one without the other.

There are scores of cultural differences between North and South California, just as there are between Newark and the Pine Barrens of South Jersey, or between dozens of other regions crowded together in the same state. Most of these differences hold no significance. There is one, though, that may bear on the question of whether to split the state.

Northern art, literature, tourism, and other cultural elements reveal a general sense that people should in some way struggle against the separation from nature that modern life is forcing upon all of us. This is evident even in restaurants. The latest in chic food in the North is a trend started by that terrific Berkeley restaurant Chez Panisse (and copied now by several others). In these establishments the food comes from gardens operated by the restaurants, or in some cases from the wild. A row developed in early 1983 when a state health official banned the use of wild chanterelle mushrooms by another famous Berkeley restaurant, the Fourth Street Grill. Owners and customers of these "real food" establishments were outraged by the decision. In some northern parts, the preparation of lovingly cultivated, carefully prepared natural foods is almost a religion. The urge involved here is the desire to get closer to the earth that sustains us, as a matter of daily life. As technology pushes us away from the earth, people in North California want not to drag their feet, but rather to fight their way back, perhaps bringing what they can of modern technology with them.

The North, however, is a hospitable land. The Southland, despite its natural beauty and great climate, is not so bounteous, not so inclined to support large human populations. This has brought about a different attitude. Technology is all that makes life possible; therefore, separation from nature is embraced, accepted, even idolized. This cultural difference between North and South might not be too hard to live with, except that all the people of California are faced with an extremely difficult problem that they must solve together: many of the people are concentrated in one place, much of the food is grown in another, and most of the water is found in yet another. The solution to this problem involves making decisions about our relationship to nature and technology, which, in turn, strikes at the philosophical Achilles' heel of California, the differing mindsets of North and South. This problem and the way it works to divide California is the subject of the next chapter.

The Two Californias Photo Quiz

Have you got the difference between Northern and Southern California completely straight in your mind? If you feel confident, test your powers of discrimination on the following photographs. Give yourself 10 points if you correctly identify the precise location of each photograph. If you can only determine whether the photograph was taken in North California or in South California, give yourself 5 points. South California is defined here as the ten counties of Imperial, San Diego, Riverside, Orange, San Bernardino, Los Angeles, Ventura, Kern, Santa Barbara, and San Luis Obispo. North California is the rest of the state. A score of 120–160 points qualifies you as a split-state expert. A score of 80–119 is average. Any score below 80 points means that you need a more thorough grounding in our differences.

1.

2.

3.

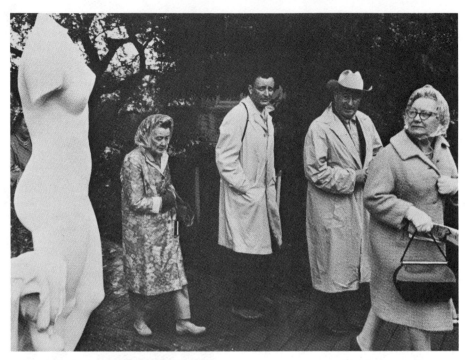

4.

5.

6.

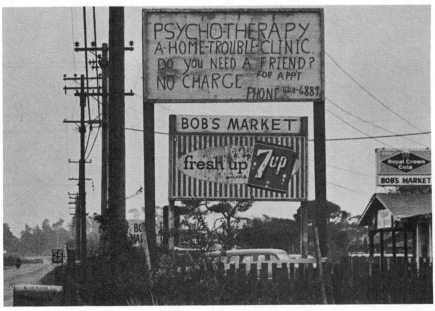

7.

8.

9.

10.

11.

12.

13.

14.

15.

16.

Answers to the
Two Californias Photo Quiz

1. Yosemite Valley, of course, North California *(Rondal Partridge)*.

2. Beach along the Great Highway, San Francisco, North California *(Richard Dawson)*.

3. Santa Monica, South California *(Baron Wolman)*.

4. Hearst Castle, San Simeon, South California *(Michael Bry)*.

5. Pulgas Water Temple, San Francisco Peninsula, North California *(Richard Brooks)*.

6. Eureka, North California *(Michael Bry)*.

7. Mono Lake, North California (U.S.G.S.)

8. Lake Arrowhead Village, South California *(Victor Stein)*.

9. Moss Landing, Monterey Bay, North California *(Baron Wolman)*.

10. Los Angeles, South California *(Baron Wolman)*.

11. Daly City, North California *(Baron Wolman)*.

12. From Mt. Palomar, near San Diego, South California *(William Aplin)*.

13. San Marcos Pass, north of Santa Barbara, South California *(Santa Barbara Chamber of Commerce)*.

14. Belvedere Lagoon, Marin County, North California *(Pete Peters)*.

15. Smith River, the northernmost town in North California *(Baron Wolman)*.

16. Imperial Beach the southernmost town in South California *(Baron Wolman)*.

Wolman photographs from *California from the Air* (Squarebooks); Bry photographs from *This California* (Diablo Press); Dawson, Brooks, Stein, Aplin, Santa Barbara, and Peters photographs from *Beautiful California* (Sunset Books)

4

Water Politics and the Third California

Imagine a mountain reservoir so enormous a person standing on one shore can barely make out the water's edge on the far shore. Imagine an electric pump so powerful it can lift thousands of pounds of water thousands of feet uphill over the tops of mountains. Now imagine a complex of canals, aqueducts, forebays, bypasses, sloughs, intakes, and concrete dams so massive it can be seen by eye from outer space.

> "A friend of mine in hydrology once described the construction of a dam as man's ultimate way of thumbing his nose at God."
> —William Kahrl

You have just envisioned California's State Water Project and the federal Central Valley Project, which stretch across hundreds of miles of California's interior valleys. Together the two systems capture and move enough water for the needs of 40 mil-

lion people and for California's number one industry, agriculture.

Water is the lifeblood of California. More than the discovery of gold and oil, the construction of the railroads and the freeways, more than the great land and people booms and the rise of the film and aerospace industries, the damming of California's streams and the construction of the state's water delivery systems have made California what it is today. The metropolitan centers of San Francisco and Los Angeles—the two largest cities in the US that import their water long distances—would not be the unique places they now are if city planners at the turn of the century had not foreseen the need to bring water from the faraway Hetch Hetchy Canyon and the Owens Valley. By the same token, the state's richest industry would not have made California the nation's premier agricultural state if early farmers had not figured out how to divert clear mountain streams to irrigate their arid fields. Today reliance on nature's beneficence and vulnerability to its vagaries are just as much a part of life for the tanned date-shake vendor in scorching Indian Wells as they are for the middle-aged marijuana farmer along the wind-battered North Coast.

While this common interest unites every Californian with every other, hassles over that water have, for decades, driven wedges of apparently irreconcilable differences between water-rich and water-poor regions. As the South swelled in population at the turn of the century, the problem of where to secure adequate water supplies intensified. By the middle of this century, water had surpassed the Gold Rush, the break-up of the ranchos, ruinous taxes, inequitable representation, and "nature's unerring hand" in marking a border as the number one reason to split the state in two.

This shift came about because of the uneven distribution of water resources in California and because of how the state developed. The northern part of the state contains 80 percent of California's naturally occurring water, while the southern third holds 60 percent of the state's population—and controls the legislature. California developed backwards, or rather upside-down—from the water-rich North to the water-starved South. For a variety of reasons, which we discussed in chapter 2, more people flocked to the arid Southland than ended up in the rainy North. While water is the well from which California's great wealth and plenitude spring, this uneven ratio of people to water has made water politics the most divisive issue in the state today.

Powerful lobbying groups have been set up to press for more dams, more canals, more levees—or, conversely, for fewer water projects, more water conservation, more realistic water pricing. In recent years, statewide initiatives on crucial water questions have triggered expensive pro and con campaigns and set off fierce battles between the "haves" and the "have-nots."

One fight, now renowned to many Californians, was that over the New Melones Dam and Reservoir on the Stanislaus, a popular whitewater rafting river in a remote section of the Sierra foothills. The federal government, in the embodiment of the Army Corps of Engineers and the Bureau of Reclamation, wanted to fill the reservoir in order to sell the former whitewater to growers in the Central Valley for irrigation. Opponents argued that the farmers don't need the Stanislaus water, the fish in the river do. A few years ago, when the Corps started to fill the reservoir, a half dozen river-lovers were moved to threaten suicide. The Corps relented, filling the reservoir only part way, but the heavy winters of 1981–3 filled it nearly to the top.

Southern California has always maintained that it needs northern water to survive. As long as the South holds the cards, it can continue to deal in water projects to overcome its awkward ratio of water to people, and there won't be much Northerners can do about it. For this reason many have argued, and still do, that instead of more northern water for the Southland, the solution to this troublesome equation is *two*, politically independent, Californias.

But wait.

There is a *third* California.

Since 1948, California has been the nation's number one agricultural state. California agriculture—more accurately known as agribusiness—generates approximately 10 percent of the US's agricultural income, $14 billion in 1981. The state's agriculture is one of the most diversified in the world. It grows some 250 crops, seeds, flowers, and ornaments. California produces more grapes than any other state and, in 1982, was first in cotton production. Because so many different types of crops are grown, each (with the exception of cotton and grapes) contributes less than 2 percent of the state's total agricultural output. Cotton accounts for 9 percent and grapes make up 8 percent.

The list of superlatives relating to California agribusiness goes on and on: number one producer of 48 of the 69 major US crops, including sugar beets, eggs, almonds, walnuts, pistachios, and strawberries; number two producer of dairy products; number three in turkey and sheep production; and so on.

More to the point, however, California agribusiness consumes at least 85 percent of all the water used in the state. So while the North and the South bicker over the Peripheral Canal and an Eel River dam, the fight over California's water cannot really be understood without first appreciating the role played by agribusiness, the Third California.

First, some definitions: *The American Heritage Dictionary* defines *agriculture* as, "the science, art, and business of cultivating the soil, producing crops, and raising livestock useful to man; farming." It defines *agribusiness* as, "farming engaged in as big business, embracing the production, processing, and distribution of farm products and the manufacture of farm machinery, equipment, and supplies."

California is unique among farming states. Although American ideology has, throughout history, viewed farming as "a way of life for independent farmers of middling means," in the words of Lawrence Jelinek, in California, large-scale growers have been the dominant pattern from nearly the beginning of the enterprise.[1] By the 1920s, "an agricultural elite composed of larger scale farmers and corporation farmers had made California the nation's first and most pervasive example of industrialized agriculture."[2]

The average size of California farms continues to grow. Farmlands are being concentrated into fewer and fewer hands. A 1978 report shows that 3.7 percent of the largest growers in the state control 59 percent of the state's farmland.[3]

One quarter of the food on America's tables comes from California. And there's plenty left over to ship overseas. The state grows 40 to 45 percent of the fresh fruits and vegetables in the US, including lettuce, processing tomatoes, lemons, peaches, apricots, olives, figs, avocados, and garlic.

California's great soils make up just one reason for the state's agricultural cornucopia. California is the only region in North America with a winter rain, summer drought cycle—characterized by mild winters, rainless summers, and long growing seasons. These are perfect conditions for growing fruits, nuts, winter vegetables, cotton, and rice.

The most important factor in the astonishing success of California agriculture is the ability to move water from where it is to where it isn't. A number of other developments—some recent, some a hundred years old—also contributed to the growth of California agriculture and helped secure its position among the top ten agricultural producing regions in the world.

Early Farming

Before farmers were able to control the flow of rivers and pipe water to their fields, the agricultural potential of the Central Valley went largely unnoticed. "In its natural condition, the valley, from the Delta to its southern terminus at the Tehachapi mountains," states the *California Water Atlas*, "was a spacious dry grasslands hundreds of miles long, a Kansas in California. Just as the grasslands of the eastern Great Plains were grazed by huge herds of buffalo, so the San Joaquin Valley had its own large animal herbivora, which roamed the flatlands by the thousands—the tule elk and the pronghorn antelope."[4]

The valley's first settlers, after trying unsuccessfully to grow crops that could withstand the long dry summers and the winter floods, turned to cattle grazing. Marc Reisner, author of the *Cadillac Desert*, a book on water in the West to be released in 1984, states, "By 1880, the Central Valley...had a population of cows [about 3 million] several times larger than California's population, fattening alongside its two great rivers, the Sacramento and the San Joaquin."[5]

The severe drought of 1863–1864 wiped out most of the herds. The valley was, as historian Robert Glass Cleland described, the scene of despair and devastation in the fall of 1863 and for years to come: "Day after day went by with cloudless skies, and the grass failed to sprout from the famished earth. The springs and waterholes dried up, and the great ranges were eaten bare of every kind of feed. . . . The loss of cattle was fearful. The plains were strewn with their carcasses. In marshy places and around the *cienegas*, where there was a vestige of green, the ground was covered with their skeletons, and the traveler for years afterward was often startled by coming suddenly on a veritable Golgotha—a place of skulls—the long horns standing out in defiant attitudes, as if protecting the fleshless bones. It is said that 30,000 head of cattle died on the Stearns Rancho [the largest cattle ranch in California] alone."[6]

Hundreds of cattle ranchers, including many of the Californio rancheros, were wiped out along with their herds. Like vultures, land speculators soon alit, hoping to profit from the ranchers' bad fortune. They bought up the devastated rangelands in the Central Valley and farther south and divided them into 40-acre tracts, which sold at prices ranging from $2 to $10 an acre.

The cheap lands brought a rush of immigrants to central California from the northern part of the state and from the East. Unaware that the area offered precarious farming conditions, the new settlers set to tilling the earth. The state government offered bounties to farmers who experimented with such exotic crops as tobacco, hemp, cotton, silk, tea, and coffee, but these crops did not flourish. Dr. Hugh J. Glenn, meanwhile, was working wonders with non-irrigated wheat in the Sacramento Valley. His success prompted others to sow their soil with the hardy grain. Wheat quickly replaced cattle as the state's leading agricultural commodity.

The long hot summers of California's interior valleys produced a hard dry wheat that could withstand the long journey to foreign markets, thereby enabling it to compete with grain from the world's leading producers. Within a few years, California became one of the chief grain-producing states in the United States. It developed an exclusive trade with Liverpool, England, which imported vast quantities of the durable wheat for distribution throughout Great Britain and guaranteed huge profits for California's growers.

In order for wheat-growing to be profitable, farmers had to plant it in large enough tracts of lands. This meant that a good deal of capital investment was required for the enterprise. During this period, "gentlemen farmers" from San Francisco and other cities bought up vast acreages in the valley. They either worked the land themselves part-time or hired others to farm it while they kept the books and arranged for international credit and trade with overseas markets. By the second half of the 1800s, most of the farm lands in the San Joaquin and Sacramento Valleys were owned by a handful of individuals and companies.

European immigrants William Miller and Charles Lux, both formerly butchers in Gold Rush San Francisco, stand out among the largest landowners. After the Gold Rush, Miller began buying rancho land in the northern San Joaquin Valley to graze cattle and to grow feed grains. While Lux took care of the business in San Francisco, Miller bought up more and more parcels of land in the valley, using both honest and devious tactics. Recognizing the agricultural potential of drained swampland, Miller bought such tracts for no more than $1.25 an acre. By swearing he had spent an equal amount reclaiming the land, he got reimbursed by the state for his purchase price. Eventually Miller and Lux owned both banks of the San Joaquin River from west of Modesto to near Madera—more than a hundred miles along the river—as well as a 50-mile strip along the Kern River. Ultimately, Miller and Lux built an empire of more than a million acres—700,000 in the San Joaquin Valley alone.

The term "Miller and Lux" came to represent, in the mid-1800s, the very idea of land monopoly and economic domination throughout the San Joaquin Valley. Espousing the philosophy of a dyed-in-the-wool speculator, Miller once stated, "Wise men buy land, fools sell."[7]

Another land speculator, William S. Chapman, arrived in California from Minnesota in the 1860s and soon became the state's largest landowner. With his partners, he acquired more than a million acres by 1871. Following the model of such early agricultural settlements as the Mormon colony at San Bernardino, Chapman established the Central Colony in the Fresno area.

The Central Colony experimented with cultivation techniques, contributing to the valley's growing production of high-quality wheat. But Chapman's greatest contribution was organizing, along with other prominent landowners, the Fresno Canal and Irrigation Company and the San Joaquin and Kings River Canal and Irrigation Company. By 1870, 60,000 acres in the Central Valley were under irrigation.

While most Central Valley growers cultivated the highly profitable wheat, other farmers grew grapes. Established at the Spanish missions and continued after the American conquest, California's wine industry had become

Hay mowing. Courtesy, the Bancroft Library

a major part of its agriculture by the 1860s. The early vineyards had been in Southern California, with Los Angeles the center of the industry in the 1830s and '40s. But superior growing conditions in the North lured enterprising vintners to the Sonoma, Napa, Livermore, Santa Clara, San Joaquin, and Sacramento valleys after the Gold Rush. Following a boom in the '60s and '70s, a tiny plant louse and a deep economic depression nearly destroyed the wine industry. The wine producers rallied and beat both the pest and the depression, and in 1900, more than 150,000 acres were producing 19 million gallons of wine—more than 80 percent of the total volume pressed in the United States. Today, grapes are California's number four agricultural product, with a 1981 income of $1.118 billion.

During California's bonanza wheat era, farm mechanization advanced rapidly. An insufficient harvest labor supply and the flat land in the Central Valley encouraged development of the combine harvester, the steam-powered tractor, and other innovations. By the 1880s, California's grain farming was the most mechanized in the world.

By the end of the nineteenth century, California's booming wheat industry had begun to bottom out. Competition from Russia, the Midwest, and other producers slashed the international price of wheat. Four decades of growing only wheat had severely depleted the valley soil's nutrients, causing steep declines in both yield and quality. And, as settlers demanded more and more land in California and early irrigation efforts revealed greater potential profits—on fewer acres—from fruits and vegetables, many wheat growers began converting or selling their land.

Those who kept their land found the valley's soils and climate well suited for fruits and vegetables. The completion of the transcontinental railroad to the West Coast in 1869 had already supplied an essential ingredient to the budding specialty crop industry: it provided the missing link between California's prodigious output and needed markets on the East Coast and abroad. Before the railroad arrived, only durable goods such as the dry wheat could be shipped any distance; now the perishable fruits and vegetables could be sent to faraway markets with a good chance of arriving intact.

Besides connecting California farms to major markets, completion of the transcontinental railroad also gave growers an unprecedented supply of cheap labor—newly unemployed Chinese railroad workers. Growers had to pay high freight costs to get their crops east, but they compensated by paying the Chinese very little for their patient field work and careful handling of the delicate new specialty crops.

Farmers Begin to Irrigate

Even more vital to the emergence of the highly profitable specialty crops was the spread of irrigation. Many of the Central Valley's farmers pooled their resources to form irrigation districts and build larger dams and canals.

Early efforts to tap California's free-flowing streams met with two major obstacles. First the old conflict between farmers and miners had to be resolved. Since the 1850s, hydraulic mining operations in the Sacramento Valley had been hosing down hillsides and letting the resultant debris—thousands of tons of gravel—tumble into the rivers below. The accumulated debris wreaked havoc on the farmlands downstream, covering them with the silt and ruining crops. Angry farmers, as we saw in chapter 1, banded together to form an Anti-Debris Association and took their case to court. In 1884 they won; a circuit court outlawed the dumping of mining debris in rivers. The decision sounded the death knell for the multi-million dollar hydraulic mining industry and signaled the growing political clout of California agriculture. The clash also triggered a vehement drive to split the state, as we have seen.

Would-be irrigators faced an even bigger stumbling block—California water law. Inherited from English common law when California entered the Union, California water law granted the primary right to the use of water in a stream to those who owned land adjacent to the stream, but stipulated that

they use it only on that land. Appropriators—those who would divert water
to use on land away from a stream—could do so, *after* the riparian owners—
landowners adjacent to the stream—had taken all they wanted. This is the
principle of prior appropriation, also known as "first in time, first in right."

During the 1880s, it became increasingly clear that a system of water law
developed in a wet and rainy climate like England's had almost no relevance
to an arid region like the Central Valley. By this time, farmers had realized
the need to irrigate their land and were acquiring the means to construct the
expensive canals and storage facilities. The time was ripe to test the controversial riparian doctrine.

During the early 1880s, the two biggest landowning partnerships in the
San Joaquin Valley got into a bitter dispute over rights to the water of the
Kern River. They argued all the way to the California Supreme Court.

Viewed as the crucial test between appropriative and riparian rights, the
case of *Lux* v. *Haggin* was fairly straightforward. Appropriators James Ben
Ali Haggin and Lloyd Tevis had dammed the Kern River and were diverting
water to irrigate their land 30 miles away. William Miller and Charles Lux,
whom we met earlier, held vast riparian rights along the river. They sued
Haggin and Tevis to stop the upstream appropriation and restore the un-
impeded flows to their land. In 1886, the Supreme Court ruled in favor of
Miller and Lux's riparian rights, but did not reject the principle of appropriation. The decision did nothing more than uphold the same practices that
had plagued California for years. It triggered an explosion of protest; angry
irrigationists, dependent on the doctrine of appropriation, successfully lobbied
the legislature to hold a special session that same year for the purpose of considering irrigation legislation. Although nothing was accomplished that
session, the following year saw the enactment of the Wright Act of 1887.

A major breakthrough for irrigationists, the Wright Act declared, for the
first time ever, that irrigation was a public use. The act provided for the
creation of public irrigation districts with authority to supersede riparian
water rights by invoking the right of eminent domain. Upon passage of the
Wright Act, dozens of irrigation districts were formed in Southern and Central California.

Irrigation and reclamation of marshlands and riparian woodlands for agriculture, by the turn of the century, had transformed the very landscape of
the Central Valley. In place of dry prairies and wetlands, "the valley was a
carpet of plenitude," Marc Reisner writes. "Anything could be grown. In the
north there were almonds and rice and beans. In the south there were oranges

and cotton and cantaloupes. In the Delta near San Francisco asparagus spears shot like green rockets from the perfect soil."[8]

California agriculture benefitted from bold experimentation with new crops and cultivating techniques, begun as early as 1860. In the latter half of the nineteenth century, several growers introduced foreign crop strains into the Central Valley, including the wildly successful zinfandel grape. With the founding of the California College of Agriculture (now the University of California at Davis) and the US Department of Agriculture Extension Service, experiments in crop genetics became a crucial part of California's agricultural development. Today, California growers don't make a move without first consulting agronomists and crop geneticists about the suitability of their land and operations to a particular crop or strain.

Rondal Partridge

Land Concentration

Between 1900 and 1920, farm colonies proliferated in California. Most were made up of small farmsteads, created out of the large wheat farms and recently subdivided. Their survival often depended on their ability to secure adequate supplies of water for irrigation. As more acres were planted, farmers began to rely on the groundwater below their land, but pumping was expensive and many small operators were driven out of business by farming's mounting costs. In their place, large-scale operations moved back in, many of them corporations. Ownership of land in the valley was falling into fewer and fewer hands. Several factors played a role, including these:

• fruit and vegetable production depended on expensive irrigation;
• marketing cooperatives, formed during this period, handled only a specific crop and worked with the processing companies to set the volume, grade, and prices at which farmers could sell crops;
• harvest labor had to be available and affordable at the right time; and
• banks often conditioned loans on a farmer's willingness to grow the crops that the bank judged to be the best suited to the farmer's land.

In addition to concentrating land ownership and forcing many small farmers off the land, these pressures encouraged farmers to specialize in one or two crops rather than to plant their fields with a variety. And as farm enterprises grew in size, they began to move into food processing and distribution. In 1916 an Italian immigrant named Mark J. Fontana succeeded in merging four large packing companies into the California Packing Company. Historian Jelinek observes: "Calpak not only owned and leased many thousands of acres, but it contracted with thousands of growers for their fruit and vegetable crops. . . . Once it had processed these crops, Calpak shipped them to its nation-wide warehouse network for distribution under its Del Monte brand name. From its inception, Calpak was the largest canning operation in the world."[9]

Not all small farmers were driven out of business by these trends. Many have survived, able to adapt and thrive on their own initiative and with help from their government. In 1902, Congress passed the Reclamation Act to thwart the growing trend toward land concentration and to assist the small farmer by providing cheap irrigation water. Large corporate growers resorted to clever legal ploys to get around the law's limit on the size of farms (160 acres) eligible for water from federally subsidized reclamation projects. Others simply refused to comply with the limit, and the government was lax in enforcement. For decades, lawsuits attempting to coerce the government into enforcing the law have raged on in the courts. Congress finally ended the fight in 1982 when it adopted amendments to the Reclamation Act that increase the acreage limitation to 960 acres and require excess land to be sold at low government-set prices or go dry. Representatives of small farmers call the compromise a defeat, since the new law eliminates the requirement that farmers live on or near their land. Although they don't share all interests with their larger, corporate brethren, most of state's small farmers go along with the wealthier, politically more powerful agribusinesses when their interests do coincide—as they do in demanding more water projects.

California Agribusiness Today

Along with their profits, California's corporate growers' political savvy has risen. In 1919, they organized the California Farm Bureau Federation, the state's largest farmer organization, with 100,000 members, and the big growers' most influential lobbyist. By the 1940s, it looked as though agribusiness could get just about anything it wanted. It convinced Congress and the USDA to bail it out with cheap reclamation water after it had depleted its underground aquifers; it continued to thumb its nose at efforts to enforce the Reclamation Act's 160-acre limit; it got all the cheap labor it needed at harvest time by relying on displaced Americans and Mexicans working illegally in the US. It even convinced Congress to legalize its use of Mexican labor during the 1940s and '50s, and to boost its sagging profits with price supports. And it spent an estimated $8 million in 1981 and 1982 alone to persuade Congress to throw out the hated 160-acre limitation.

In spite of its considerable political clout, modern agriculture in California has come under attack of late, for reasons beyond its control. Increasing air pollution, blown in from the cities and trapped by the mountain ranges on either side of the valley, along with skyrocketing energy costs and the loss of prime cropland to urban sprawl, threaten California's bountiful harvests. The growers' response has been, in large part, to turn to their friends at UC Davis and USDA Extension Service. In these facilities, agronomists are crossing wild Ethiopian and Peruvian strains of corn and wheat with their domesticated California relatives and splicing genes from particularly hardy plants with genes of other plants that have different desirable traits. They want to create hybrids that can withstand deteriorating air quality and increasing soil salinity (which comes from pesticides, fertilizers, and salty irrigation water);

Rondal Partridge

use less water; stand up to mechanical harvesters (because farm labor costs have risen as farmworkers have organized); and still maintain high yields per acre.

California's agriculture industry is hardly on its last legs. Just as it has so many times in the past, it will overcome its present problems, even if some land has to be taken out of production or converted to different crops. A $14-billion industry, born of hardy people and dreams, doesn't collapse overnight. Indeed, as long as the Third California maintains its political alliances with such traditional pals as the Metropolitan Water District of Southern California and is able to throw its weight around in Sacramento and Washington, it will be truly formidable.

Meanwhile, on the Coast

While farmers switched from a wheat monoculture to diversified fruits and vegetables, the metropolitan centers of California grew by leaps and bounds. As settlers from the East Coast and Midwest poured into the state, the water sources that Los Angeles and San Francisco had relied upon for decades began drying up.

"The history of California in the twentieth century is the story of a state inventing itself with water," states William Kahrl in *Water and Power: The Conflict over Los Angeles' Water Supply in the Owens Valley*.[10] During the first part of the state's development, early settlers chose areas near sufficient water supplies. The Los Angeles pueblo, in fact, was located in the basin of the Los Angeles River; the Franciscan missionaries assumed the river had abundant water. Then the Gold Rush brought settlers by the thousands to Northern California and inundated its then-largest city with '49ers who stayed. Soon after, the railroads and their wily promoters lured tens of thousands to the arid Southland.

Kahrl continues: "Prosperity and development could not be assured without additional sources of [water] supply. Thus, confronting a common problem but acting independently and exclusively in their own interests, San Francisco and Los Angeles set out simultaneously to develop distant watersheds in a race that would ultimately go a long way toward determining which city enjoyed supremacy among the commercial centers of the Pacific Coast." Though this last part is debatable, it is true that the two cities were forced to come up with creative solutions to their water shortages, due largely to California's loyalty to the riparian doctrine. This English common law holdover gave the communities of the North Coast and the cities situated on the great rivers of the interior a natural and legal advantage in controlling California's water supplies.

So what did San Francisco and Los Angeles, coastal communities far from the state's great rivers, do? The northern city twisted the federal government's arm and got a permit to use a water supply lying in the public domain, the stunning Hetch Hetchy Canyon in Yosemite National Park. Los Angeles, taking an altogether different tack, deceived the ranchers and farmers in the Owens Valley, embarrassed the federal government into backing out of its plans for the valley, and bought up riparian rights along the Owens River.

"The construction of the Hetch Hetchy and Los Angeles aqueducts marks the true beginning of the modern water system of California," Kahrl writes.

They are significant not only because of their huge scale and cost, but because these projects were the first instances of water being moved long distances where "the water itself was the principal object of the enterprise."

"No Holier Temple"

As early as 1882, city planners in San Francisco began to consider new ways to supply water to the still booming city by the bay. One of the most promising schemes involved carrying water from the upper Tuolumne River in the Sierra Nevada west to San Francisco. The reservoir site, the narrow Hetch Hetchy Canyon, was viewed as ideal because of its steep walls. The city didn't act until 1901, when the city engineer recommended that San Francisco acquire water rights along the Tuolumne and build its own dams and reservoirs. It had, until then, relied on the services of a private utility, Spring Valley Water Company, which had come under serious charges of corruption and price gouging.

Although the proposed site was in Yosemite National Park and thus considered inviolable, the city requested a formal permit from the Department of the Interior to develop Hetch Hetchy for its water. In 1903, Secretary of the Interior E. A. Hitchcock rejected the city's application, igniting a ten-year battle between dam proponents and opponents. San Francisco Mayor James Phelan led the campaign to get the project approved and enlisted the support of Gifford Pinchot, the country's leading forester, along with President Theodore Roosevelt, an outspoken preservationist.

The proposal to dam Hetch Hetchy pitted the two leading spokesmen for the country's budding conservation movement against each other. Pinchot advocated planned and regulated use of the public's resources; Sierra Club founder John Muir found the entire proposal abhorrent. "Dam Hetch Hetchy?" he cried. "As well dam for water tanks the people's cathedrals and churches; for no holier temple has ever been consecrated to the heart of man."

Besides Muir and the Sierra Club, foes of the plan included the Spring Valley Water Company, which did not welcome competition from a city-owned utility, and another utility, the Bay Cities Water Company, which wanted San Francisco's business.

In 1906, the great earthquake struck San Francisco, rupturing the Spring Valley pipeline and leaving the city without enough water to fight the fires that broke out. Seizing the moment, Bay Cities proposed its own water delivery plan to civic leaders. Claiming its plan would be cheaper than the Hetch Hetchy project, Bay Cities offered to sell San Francisco, for $10.5 million, its water rights on the American and Cosumnes rivers near Lake Tahoe. Then it came out that the city's political boss, Abraham Ruef, was to receive $1 million of that in "attorney's fees," to be divided among himself, the mayor, and the supervisors. The scheme collapsed. The city turned its attention back to Hetch Hetchy.

In 1908, Roosevelt's Secretary of the Interior, James R. Garfield, a Pinchot supporter, reversed his predecessor's decision and granted San Francisco a permit to develop Hetch Hetchy.

The permit enraged John Muir and the expanding legion of conservationists around the country. They set out to get the permit revoked. Muir and others wrote articles on the importance of preserving wilderness, both for what it is and for what benefits can be derived from it. In the magazine *Outlook*, Muir wrote: "Everyone needs beauty as well as bread, places to play in and pray in where nature may heal and cheer and give strength to body and soul alike." The preservationists argued that Hetch Hetchy was a popular recreation area for both the physically active and the contemplative visitor. Taking advantage of the nation's sensitivity to charges of American commercialism, Muir labeled the would-be developers "temple destroyers" who rejected "God of the Mountains" in pursuit of the "Almighty Dollar."

Such eloquence struck a chord in the hearts of many Americans. When the Howard Taft administration took office, the new Interior Secretary rescinded San Francisco's permit. But the project's backers persevered and finally won. In 1913, Congress passed the Raker Act, granting the City of San Francisco the Hetch Hetchy Canyon. The city then began construction of a reservoir and an aqueduct stretching 186 miles across the state for the sole purpose of bringing Yosemite Park water to the Bay Area. Construction took years; the first water from the Upper Tuolumne did not reach San Francisco until 1934.

"There It Is! Take It!"

While the North suffered tension headaches over water, Southern California got migraines just thinking about it. Los Angeles, remember, is primarily a desert or semi-desert—"the land of the *paisano* or road runner, the horned toad, the turkey buzzard," in Carey McWilliams' words.[12] The Los Angeles Basin has 1,400 acres of habitable land, 6 percent of the state's total, but only .06 percent of the natural stream flow of water in California.

Los Angeles' traditional source of water was the Los Angeles River, a pokey, shallow stream whose tributaries in the San Gabriel, Santa Monica, and Santa Susanna mountains pour their meager flows into vast underground reservoirs in the San Fernando Valley. The Spanish government had granted full use rights to the river's flows to the pueblo of Los Angeles at its founding in 1781, but competitors began to assert their rights to the river's water during the 1870s. Irrigators in the San Fernando Valley, in particular, claimed riparian rights to the underground aquifer. In 1881, the California Supreme Court affirmed the city's historic claim to the river's entire flow.

As early as 1899, the city's burgeoning population was placing undue demands on the Los Angeles' meager flows. The city began to sink wells to tap its underground reservoirs. By 1903, after three particularly dry seasons, enough wells and tunnels had been dug to irrigate the entire state of Rhode Island. But still, the city needed more.

Urban leaders predicted continuing population growth and sought to get their hands on a plentiful supply for the city's growing needs. The search was led by two powerful men, Fred Eaton, a former mayor of Los Angeles and an engineer, and William Mulholland, superintendent of the Los Angeles Department of Water and Power.

The son of a prominent South Coast engineer, Fred Eaton was "youthful, aggressive, innovative, startlingly handsome, and possessed of a charm that would win him the respect and even the affection of his adversaries throughout his life," William Kahrl writes. He eventually became mayor and led the drive to regain public control of the municipal water system, which at one point the city had relinquished to the Los Angeles City Water Company, a private concern. After years of haggling, the city succeeded, in 1902, in buying it back. Along with the company's waterworks and facilities, Los Angeles acquired, again in Kahrl's words, "an asset that proved to be of far more enduring value." That asset was the company's supervisor, William Mulholland.[13]

Born in Ireland to a modest middle-class family, Mulholland was cut from a different cloth than was Eaton. He came to the United States in 1874 as a sailor, then traveled from New York to the Great Lakes to Pennsylvania, doing odd jobs. While in Pittsburgh, working at an uncle's dry goods store, young Mulholland came across Charles Nordhoff's accounts of California and decided to head west.

He ended up in Los Angeles and took a job on a well-drilling rig. "The first well I worked on changed the whole course of my life," Mulholland wrote later. "When we were down about 600 feet, we struck a tree. A little further we got fossil remains and these things fired my curiosity. I wanted to know how they got there, so I got hold of Joseph Le Conte's book on the geology of this country. Right there I decided to become an engineer."[14]

After a brief stint of prospecting in the Arizona Territory, Mulholland returned to Los Angeles, a place that was to hold his lifelong affection. Recalling
later in life his early experiences in the City of the Angels, Mulholland wrote:
"Los Angeles was a place after my own heart. It was the most attractive town
I had ever seen. The people were hospitable. There was plenty to do and a
fair compensation offered for whatever you did. . . . The Los Angeles River
was the greatest attraction. It was a beautiful, limpid little stream with willows
on its banks. . . . It was so attractive to me that it at once became something
about which my whole scheme of life was woven. I loved it so much."[15]

He took a job as a ditch tender for the Los Angeles City Water Company
and spent hours studying the river's banks and dreaming of ways to fashion
its uncertain flows to build Los Angeles into a great city. A large, angular,
raw-boned man with a walrus moustache, Mulholland moved up quickly
through the company ranks. After only eight years, "this unschooled and un-
trained former laborer had emerged as Fred Eaton's protegé and heir apparent
for the superintendency of the company."[16]

The two of them made an unlikely pair. Eaton was elegant, well-born, and
refined. Mulholland was a gruff and blunt man who loved games and jokes.
While Eaton craved public attention, Mulholland shunned it, preferring to
spend his evenings reading for his own edification.

As company superintendent, Mulholland made the transition from private
to public employ without flinching. He proceeded to improve the city's waterworks, modernizing the entire system, improving measurement techniques,
and always looking to the future. It was Fred Eaton, however, who first saw
that Los Angeles' future lay some 230 miles to the east. Ever since private surveys in the late 1800s showed the fertile Owens Valley's potential as a source
of Los Angeles water supplies, Eaton had been promoting his own scheme to
bring the valley's water to the city.

No one listened to Eaton; most people viewed his proposal as infeasible and
surely too expensive. The plan languished until just after the turn of the century, when the newly created US Reclamation Service (later the Bureau of
Reclamation) decided the Owens Valley was an ideal spot for one of its first
projects—a dam and reservoir that would provide federally subsidized irrigation water to the valley's farmers.

Working secretly to prevent land speculators from going into the valley
and driving up land values, the Service withdrew key parcels for its project.
Eventually word got out about the federal government's plan. Owens Valley
residents were delighted; they welcomed the idea of federally subsidized
irrigation water.

While the feds proceeded to make geological and hydrological studies, the enterprising Fred Eaton was lining up support for his plan to build an aqueduct to carry Owens River water to Los Angeles. He finally convinced Bill Mulholland to pursue the idea, and Mulholland persuaded the city water commission to conduct its own studies.

Without official permission from either the federal government or the city, Eaton then went to Owens Valley to buy up options on land alongside the Owens River. Calling himself an agent for the Reclamation Service, he was greeted with open arms. But he was really laying the groundwork for his own scheme, from which he stood to profit enormously.

Shortly thereafter, the Reclamation Service, perplexed by Eaton and Mulholland's activities, announced its decision to abandon its Owens Valley project. The news exposed Eaton's duplicity and enraged Owens Valley residents. The citizens of Los Angeles responded to the news of a planned aqueduct with mixed feelings. They hadn't seen a need for more water, and the project was certain to cost them dearly.

Now too far in to give it up, Mulholland found himself in the unhappy position of needing to muster popular support for the aqueduct. A Los Angeles municipal vote on $25 million in bonds to fund the project was scheduled for September 1905.

The city's newspapers took up positions of varying support for the aqueduct. The *Los Angeles Times* supported the project, running enthusiastic reports during the campaign for the bond election. The *Times*-owned *Herald* and the *Express* also favored the project, praising the foresight of the city's leaders. But the new paper in town, the *Los Angeles Examiner,* fancied itself a muckraker; its writers quickly became suspicious of the city's grandiose plans and attacked Mulholland and Eaton's claims that Los Angeles needed the water from Owens Valley.

In response, Mulholland manipulated rainfall statistics and publicly exaggerated the city's water needs. In its zeal, the *Times* fabricated a story about a drought to back up Mulholland's claims that the city's supplies were dangerously low. Even more zealous supporters dumped water out of city reservoirs at night to lend authenticity to the purported shortage.[17]

The muckraking *Examiner* soon found what it was looking for. Just two and a half weeks before the election, the paper revealed a link between a huge land development scheme in the San Fernando Valley and the planned Los Angeles Aqueduct. A syndicate—composed of Los Angeles' most prominent businessmen and civic leaders, including Moses Sherman, one of the city's water commissioners—had bought up 16,000 acres of land in the San Fernando

Valley, at $35 an acre. With 1000 shares each in the San Fernando Mission Land Company, these investors stood to gain millions once water arrived from Owens Valley and was put to use irrigating their arid but fertile lands.

The syndicate was led by Henry Huntington, nephew of Southern Pacific giant Collis Huntington and owner of the Pacific Electric and Power Company and the municipal railway, which had recently extended a line into the San Fernando Valley. Other members included E. H. Harriman, president of the Union Pacific Railroad; Joseph Sartori of Security Trust and Savings Bank; and the owners of the three top newspapers in Los Angeles, E. T. Earl of the *Express* and Harry Chandler and Harrison Gray Otis, publishers of both the *Times* and the *Herald*.

The scandal, however, barely fazed Los Angeles voters. Swayed by the fake water scare, the city's residents approved the bond issue by a margin of 14 to one. Begun in 1908, the Owens Valley aqueduct was to become an engineering marvel; over 250 miles long, it begins at the lower end of the valley and snakes over the mountains and across the desert to reach three large reservoirs north of Los Angeles.

As its gates opened on November 5, 1913, the aqueduct's chief engineer, William Mulholland, exclaimed, "There it is! Take it!"

As soon as it began operation, the aqueduct delivered four times as much water as Los Angeles needed or could possibly use for domestic purposes. In order to get federal approval for the aqueduct's right-of-way across Yosemite National Park, the city agreed not to sell its Owens water outside the city limits. Then it proceeded to annex much of the San Fernando Valley (including all of the infamous syndicate's holdings). Previously barren land soon "blossomed into citrus groves, beans, and potato fields," Kahrl writes, and the aqueduct, "as an urban water development project functioned for its first years of operation principally for the benefit of agriculture." Because of its annexations, the city's population quadrupled by 1923.[18]

The Owens Valley War

Within a few years, Owens Valley farmers—who instead of irrigation water got ripped off—saw their green fields wither and their crops die as the Owens River was sucked out of the valley and into the hated aqueduct. Not only had their water been taken to serve a distant city's *future* needs, it had gone first to irrigate farm land in another valley. In retaliation, the valley's farmers took up guerrilla warfare. They dynamited the aqueduct and its dams 17 times. In

Building the Los Angles Aqueduct. Courtesy, the Bancroft Library

the 1920s, when the city wanted more land and water rights, Owens Valley ranchers and businessmen demanded the highest prices for their land. When the city refused to pay such prices, residents responded with fists, guns, and dynamite. In 1924, a group of farmers seized a Department of Water and Power facility and held it for several days. In 1927, the bombings resumed and the city sent in bands of detectives and guards armed with machine guns to protect its property. Los Angeles ended up purchasing virtually all of the private lands in the valley. Today the City of Los Angeles is the principal land-owner and taxpayer in the County of Inyo, and the valley's hostility toward Los Angeles has not subsided. The city continues to suck more and more water from Owens Valley and is currently entangled in lawsuits over its right to pump and use the valley's groundwater. As recently as 1976, the aqueduct was bombed again.

The story of the Los Angeles Aqueduct is important for many reasons. It was the first time in California history that an area's water supply was drained for the benefit of a distant city. As a result, the legislature passed the "County of Origin" law in 1931 to prohibit the draining of one area's water supply for the sake of another.

What is more, it was the first time city leaders, citing a contrived water shortage, duped voters into passing an expensive water project. This tactic worked so well in 1905 that it continues to be popular with the "concrete worshippers," as water developers have been labeled.

The most important lesson to be learned from the Owens Valley tale is this: when a powerful group of people takes water from where it is to where it ain't, the people whose supply it was are not happy. Depending on how much is left for the latter group's use, they are apt to resort to extreme measures to hold onto what nature has placed in their backyard.

Within ten years after the Owens Valley aqueduct was completed, Los Angeles' booming population forced its leaders to look for more water. Historians Beck and Williams assert that it was "inevitable that water-starved Southern California would ultimately turn to the Colorado River, the one stream in all of the Southwest worthy of being termed a river."[19]

In 1928, a coalition of South Coast communities, led by Los Angeles, formed the Metropolitan Water District of Southern California to press for development of the Colorado River. In the early '30s, MWD joined hands with the Imperial Irrigation District, in southeastern San Diego County, to lobby for construction of the Hoover Dam and aqueducts to bring Colorado River water ·to Los Angeles and the rest of the dry Southland.

The alliance proved unbeatable. Imperial's farmers had been unable, until then, to convince Congress to part with the necessary funds to build the dam. With the communities' support, advocates were now "able to argue for their project on the grounds of the greatest good for the greatest number," as Peter Wiley and Bob Gottlieb point out in *Empires in the Sun*. They could, "at the same time, quietly lay the groundwork for the development of irrigated agriculture in the Imperial Valley," formerly known as the Colorado Desert. Thus, a symbiotic relationship was born: Imperial's growers would feed the growing population of Los Angeles and, in return, become part of the regional power structure, "a corporate extension of the metropolis."[20]

Congress approved the project, and the Hoover Dam was finished in 1935. By 1941, the 242-mile Colorado Aqueduct was completed; with the All-American and Coachella Canals, what would ultimately reach 5.3 million acre-feet of Colorado River water began to flow to Southern California.* (About

---

*One acre-foot equals 43,560 cubic feet.

80 percent now goes to irrigate fields in the Imperial and Coachella Valleys and the rest flows into the MWD's vast network of reservoirs and aqueducts for municipal distribution.) The arrival of Colorado River water in the Southland launched yet another round of mind-boggling growth. Southern California's population jumped from 3.3 million in 1940 to 14.8 million by 1980.

Two years after Congress approved the Hoover Dam project, Los Angeles reached out for yet more water. In 1930, the Los Angeles Department of Water and Power began construction of a 105-mile extension of its Owens Valley aqueduct, reaching north to Mono Basin, on the east side of Yosemite National Park. Within ten years, the city was diverting fresh water from Mono Lake's feeder streams and sending it south. Unable to quench its thirst, the city started taking nearly all of the fresh water from Mono's streams in 1971. The level of the once-stunning lake has dropped radically since LADWP first started taking water from the basin. Many biologists fear that the lake's ecosystem will collapse if diversions continue at their current rate.

Farmers Lobby for a State Water Project

Meanwhile, Central Valley farmers were suffering from nature's fickleness. A big drought from 1928 to 1935 forced them to drain their underground basins in an effort to save their crops. They lost a tremendous number anyway. Farmers were thus convinced they needed a big water project to keep their fields irrigated through both wet and dry years.

In 1931, State Engineer Edward Hyatt proposed an interbasin water conveyance system. His plan led to approval of the massive Central Valley Project, designed to bring Sacramento River water through the Delta and south to the San Joaquin Valley. The state was unable to sell the bonds needed to fund the project, however, and it languished until President Franklin Roosevelt came to the rescue in 1935. He approved emergency federal funds to complete the system and turned it over to the Bureau of Reclamation as one of its reclamation projects.

Over the next 25 years, the Bureau built the huge Shasta Dam on the Sacramento River, with a capacity of 4.5 million acre-feet; the Trinity Dam on the Trinity River; the Folsom Dam on the American River; and the Friant Dam on the San Joaquin River. The first federally subsidized water from Shasta Dam reached the San Joaquin Valley in 1951.

From the outset, the valley's large growers detested the Bureau's limitation of 160 acres as the maximum size of a farm eligible for the CVP's cheap water. Their unhappiness with the rule, along with a steady influx of people to California after World War II, sparked renewed interest in a state-controlled water project—what the CVP had originally been intended to be. In 1945 the California legislature created the State Water Resources Control Board to study and develop plans for a state water project and, a few years later, the Department of Water Resources was set up to coordinate the activities of all the state agencies that dealt with water.

In the early 1950s, State Engineer A. D. Edmonston came up with a state-wide plan to divert water from the Feather River to a multi-purpose dam, reservoir, and power facility near Oroville, in Butte County. The project would control floods, increase the dry-weather flows to the Sacramento-San Joaquin Delta, and provide supplies for a state-constructed delivery system to carry water from the Delta to parts of the San Francisco Bay Area, to farmlands in the Central Valley, and to the communities of Southern California.

Most Northern Californians didn't like the idea of exporting their water south, and some Metropolitan Water District directors were also wary of the proposed project. They feared that the legislature might change its mind in a dry year and rescind the District's contracts. So MWD demanded a state constitutional amendment to guarantee its water deliveries.

As a compromise, Governor Edmund G. Brown, Sr., recommended writing water delivery guarantees into a bond measure to be submitted to the state's voters. Though MWD still opposed the project, the State Water Resources Development Bond Act was ratified by the California Legislature in 1959, subject to approval by the state's electorate. The bill, known as the Burns-Porter Act, authorized $1.75 million to fund construction of the Feather River facilities and those in the Delta; provided for future claims on Northern California rivers; and approved a drain to remove agricultural wastewater from the Central Valley and dump it in the Sacramento River near the Delta.

With Burns-Porter, Pat Brown tried to please everyone. To appease the North, the act specifically guaranteed protection of water rights in the areas of origin and provided for construction of local projects. For Southern California, it required that the state not impair contracts for the sale and delivery of project water during the lifetime of the bonds.

Nevertheless, the campaign to approve the bonds in 1960 provoked one of the most divisive and hostile battles ever waged over California water. Proponents of the 1960 bond measure insisted that the water was needed for the state's growing cities and to replenish overdrafted groundwater basins in the Central Valley. Northern Californians vehemently protested the shipment of water south for the benefit of either the farmers or the cities. They didn't see the need for or the desirability of Los Angeles continuing its sprawl at *their* expense. (These arguments would be used to fight later water projects.)

Logically, the most enthusiastic support for the State Water Project came from the Central Valley farmers who were anxious to get their hands on a new supply of cheap water that was not restricted by the federal 160-acre rule.

Still unswayed by the promised written guarantees, the MWD drew up its own plan to bring water from the Eel River to Southern California. But other South Coast communities, with less secure supplies than those of Los Angeles, endorsed the state's project individually and put pressure on MWD to follow suit. Four days before the election, MWD's board of directors reversed its earlier position and signed a contract with the state for the delivery of 1.5 million acre-feet of project water. The bond measure squeaked by on November 8, 1960, due largely to the enthusiastic support of Southern Californians. The vote reflected powerful regional loyalties; throughout Northern California, only one county passed the measure—Butte, where the Oroville Dam was to be built.

Cheap Water + New Land = Big Money

Although it was designed to replenish the depleted underground aquifers in the Central Valley, the State Water Project actually served to bring new lands into production. A number of large corporations, foreseeing the enormous profits to be made once state water began flowing into previously unwatered areas of the valley, bought up huge tracts of land and installed irrigation systems. "The Tejon Ranch Company had its [irrigation] equipment in place when the first water came through because they had the capital to do so," says Donald Villarejo of the California Institute for Rural Studies.

The institute conducted a study entitled "New Lands for Agriculture," which found that on the west side of the San Joaquin Valley, primarily in Kern County, which has no underground supplies, "about 250,000 acres have been placed in production as a direct result of State Water Project deliveries."

More than 227,000 acres in this part of the valley, known as the Westlands Water District, are owned by Chevron USA Inc., Tejon Ranch Company, Getty Oil Company, Shell Oil Company, McCarthy Joint Venture A, Blackwell Land Company, Tenneco West Inc., and Southern Pacific Land Company.

"The big corporations grow permanent, high-cash crops: almonds, pistachios, olives, and grapes, not exactly dinner table fare," stated Thomas Schroeter, a Bakersfield attorney and former member of the defunct Kern County Planning Commission.

These corporations were attracted to the Central Valley not out of a love of farming but because of the tax benefits afforded to them if they grew the permanent crops. During the 1960s, the Internal Revenue Service allowed investors to immediately write off their entire share of development costs for growing almonds and other permanent crops.

Some of these corporations bought up land in the Central Valley because of the vast oil deposits lying beneath many of the now-rich croplands. Since this oil is a thick and viscous crude that must be mined with steam, extractors need large volumes of water to develop it. Though they haven't done so on a significant level yet, the petroleum giants of the Central Valley may soon be able to double their money by drilling and processing the black gold below

The Largest Landowners in Five State Water Project Districts 1980 – 1981

LAND OWNER	ORCHARDS	VINEYARDS	FIELD OR ROW CROPS	GRAZING	UNDE-VELOPED	TOTAL
			(land area in acres)[1]			
Chevron USA, Inc.	—	—	29,702	1,124	6,967	37,793
Tejon Ranch Company[2]	5,274	7,251	1,770	20,434	1,168	35,897
Getty Oil Company[3]	2,412	—	21,638	990	10,344	35,384
Shell Oil Company[4]	4,498	—	18,272	2,472	6,753	31,995
McCarthy Joint Venture A	16,105	—	7,667	1,333	—	25,105
Blackwell Land Company[5]	9,453	3,850	6,623	804	3,933	24,663
Tenneco West, Inc.[6]	4,232	831	12,896	1,109	1,112	20,180
Southern Pacific Land Co.	—	796	11,335	2,911	1,486	16,528
SUBTOTAL: 8 OWNERS	41,974	12,728	109,903	31,177	31,763	227,545
TOTAL: ALL OWNERS[7]	58,963	17,185	205,377	43,526	59,048	384,099
EIGHT LARGEST LAND OWNERS AS PERCENTAGE OF ALL LAND OWNERS	71%	74%	54%	72%	54%	59%

[1]Includes all 2,435 parcels of land, 20 acres or greater, in the following districts: Belridge Water Storage District (Kern), Berrneda Mesa Water District (Kern), Dudley Ridge Water District (Kings), Lost Hills Water District (Kern) and Wheeler Ridge-Maricopa Water Storage District (Kern).

[2]Includes Tejon Agricultural Partners

[3]Includes Getty Refining and Marketing Company

[4]Includes Kernridge Oil Company

[5]Includes the joint ventures, Hanwell Orchard and El Vic Farm Corporation

[6]Includes Tenneco Oil Company and Tenneco Realty Development Company

[7]Refers to owners of parcels of 20 acres or greater

Source: Villarejo, Don. *New Lands for Agriculture: The California State Water Project*. Davis: California Institute for Rural Studies, 1981.

their fields while they grow the highly profitable cash crops on the surface—both with the aid of cheap state project water.

Schroeter points out that the oil companies and other big corporations in Kern County "have no long-term commitment to agriculture. Blackwell Land Company [which has 13,000 acres of permanent crops and uses 42,000 acre-feet of SWP water a year] is not an agricultural company. Nor is Tejon Ranch Company, nor Tenneco, nor Shell, nor Prudential Insurance (which owns 75 percent of McCarthy Joint Venture A), nor are most of the others. They're land developers."[21]

The Tejon Company has filed draft plans with the county to build a new community of 700,000 people, and other corporations have long-range plans for residential developments on their land, according to Schroeter and Villarejo.

What's most interesting about all this corporate interest in Kern County agriculture is how it translates into the Barnum and Bailey arena of California water politics.

Ever since the Metropolitan Water District of Southern California helped tip the scales in favor of the State Water Project, it has teamed up with Central Valley growers to press for more water development. The happy alliance is based largely on a surplus of SWP water, which the MWD is entitled to but does not use. The surplus, which MWD must pay for in order to maintain its right to the water, in case the district ever does need it, goes to Kern County's farmers at wildly reduced rates—about $3 to $4 an acre-foot. Although the state is obligated by law to sell any surplus SWP water to the highest bidder, it has never done so. Explains National Land for People board member David Nesmith, "They've taken it and sold it at the cheapest cost they could [just] to pay for transportation."

Because MWD gets one-third of its income from property taxes in its vast service area, it is the urban water user who ends up subsidizing Kern's farmers, many of them wealthy corporations. (See chapter 6 for more discussion of water subsidies.)

Water for Development

So agribusiness gets cheap water for irrigation and future land developments, as well as crucial Southern California votes, out of its cozy relationship with the Metropolitan Water District. What does MWD get in return?

Bob Gottlieb, MWD director from Santa Monica—the only one who opposed the Peripheral Canal—offers one possible explanation: "MWD has

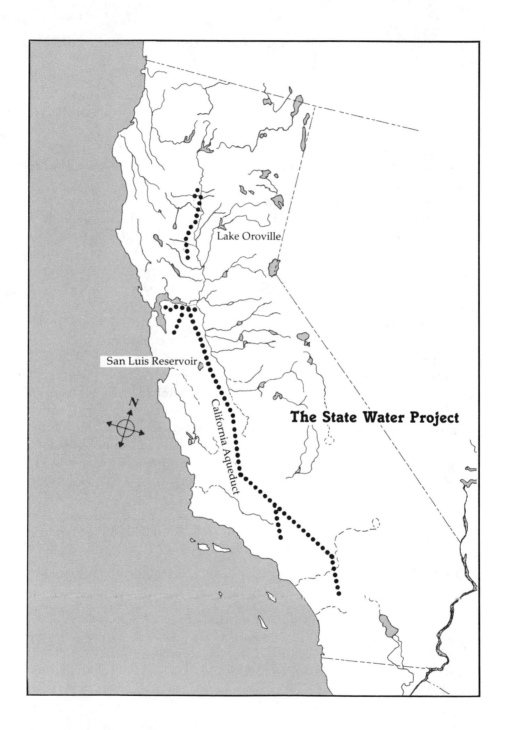

operated over the years with the assumption that the real power in the state is agribusiness, and it is constantly attempting to work out alliances and cut deals so there can be a unified power base."[22]

During the Peripheral Canal campaign, MWD was accused of acting contrary to the interests of its customers. While Met officials predicted a dusty apocalypse for the Southland if the canal were not built, Tom Graff, general counsel for the Environmental Defense Fund, charged that the MWD board had agreed to allow agricultural water users—rather than its own urban customers—to have first crack at any available State Water Project supplies in the event of a drought. (By law, urban users are entitled to what supplies there are, and farmers must be first to reduce their take during a shortage.)

"In order to keep Kern County agribusiness as part of its coalition to authorize the Peripheral Canal, MWD has agreed to short its own customers," Graff said, citing a December 1980 letter from MWD chair Earle C. Blais to the head of the Kern County Water Agency. "What it comes down to is that MWD is trading away water use priorities for political advantage."[23]

According to the conspiracy theory of California water politics, the Met wants another big water project so that it can hold on to a healthy surplus. Why? Because a surplus allows the district to achieve many things. It maintains its alliance with agribusiness, as we've seen. It fuels development in arid regions of Southern California. And it holds onto its rights to water it might actually need in the future, under the policy of "use it or lose it."

Let's consider point two, how MWD encourages desert development. "The Met has broad powers to finance or acquire any water rights deemed in the best interests of the district," *San Francisco Chronicle* reporter Stephen Magagnini reported in May 1982. He pointed out that the district has grown from a consortium of 13 cities to an agency that encompasses 141 cities and 5100 square miles and delivers water to more than half the people of California.

"The Met has nourished the growth of agribusiness and suburbia by charging low annexation fees to communities that wish to join the district and then selling them cheap water—discounts subsidized by the city of Los Angeles and other senior members of the MWD network," such as San Marino.[24]

Magagnini cites the example of a new project in rural southeastern San Diego County. Before the election on the Peripheral Canal, the MWD board tentatively approved annexation of the Honey Springs Project, 389 houses on 2100 acres near the quiet desert town of Jumul. Without MWD water, the project could not be built.

According to Dr. Gerard Petrone, president of the Honey Springs Homeowners Association, the MWD board agreed to postpone its final decision on the controversial development until after the June '82 election. When California voters turned thumbs down on the Peripheral Canal, apparently adding to MWD's headaches over water, the board approved the annexation of Honey Springs, anyway.

Petrone criticizes the MWD's decision and says, "[The developers] are wasting water here," referring in particular to an 18-acre decorative lake planned for the center of the project. He and the other landowners in the Homeowners Association are trying to block construction of the project, which Petrone calls "a Levittown in the desert."[25]

"[MWD] is being two-faced in asking the public to conserve water, while what they're really saying is anybody who wants water can have it. They're doing it for their own selfish, empire-building interests."[26]

While MWD board members deny such charges, their actions often speak louder than their words. For example, the district's desire to hold onto its rights to lots of SWP water looks innocent enough. But some people would argue that it is not merely the district's wish to have adequate supplies on hand in case of drought that motivates its desire for a surplus. These conspiracy theorists say that more important is the Met's preference for clear Northern California water over the saltier Colorado River water. If MWD keeps expanding its service area and selling a lot of water, the board can argue more convincingly that the Southland is growing and needs a more secure and larger supply of water—such as that which would be shipped south if a Peripheral Canal or through-Delta channel were built.

There are a number of other reasons the Met likes to annex subdivisions in the desert and sell them cheap water. Petrone says, "Metro has historically been on the side of developers [in order] to expand its revenue base. It is in the business of selling water, after all."

Another reason might be to promote the corporate interests of its board members. Tom De Vries and George Baker wrote in a 1980 *New West* article, "Nearly half the directors [on the MWD Board] are real estate developers or substantial landholders. Half a dozen own engineering or construction firms and 15 own or have major interests in banks and savings and loan associations."[27] They point to Preston Hotchkiss, 89, a director from San Marino since 1975, who also chairs the Bixby Ranch Company. Worth over $50 million, the Bixby Company owns property in Los Angeles, Orange, and Santa Barbara counties and is active in residential and light commercial developments around the Southland.

"The curious thing, really, is that the city of San Marino...has almost no historical interest in Met water," DeVries and Baker point out. "Since 1928 San Marino has purchased only 68 acre feet. The people of San Marino are paying taxes for State Water Project facilities they have never needed, financing water delivered to someone else [such as Honey Springs], and are being represented in this by a man for whom they did not vote [board members are appointed by the community or water district they represent], and whose business dealings suggest concerns much larger than San Marino."[28]

This reiterates what the conspiracy theorists would say about Met wanting to maintain a surplus for reasons other than the future water needs of their own tax-paying customers.

The Demon Ditch

By 1967, the Oroville Dam was finished, backing up 3.5 million acre-feet of Feather River water at Lake Oroville. The first SWP water made its way over the Tehachapis into Southern California in 1971.

Like the federal Central Valley Project, the State Water Project sends its water down the Sacramento River to the Delta. There gigantic pumps draw water out of the Delta near Tracy and send it south along the California Aqueduct, or else into one of several canals that serve Napa County and communities along the southern end of San Francisco Bay. Water destined for Southern California travels through the San Joaquin Valley until just south of Bakersfield.

There "the water enters a very different world, one that uses energy as if it hadn't gone out of style," write George Baker and Tom DeVries. "Every year enough water to put a city the size of San Francisco under 18 feet of water is pumped over a range higher than the highest falls in North America."[29]

By 1968, the California Department of Water Resources had contracted with water agencies and irrigation districts around the state to deliver 4,230,000 acre-feet of water a year. (The MWD and Kern County Water Agency are entitled to the lion's share of SWP water—over 3 million acre-feet a year combined.) The contracts provide for increasing amounts of water each year until 1990, to allow for expected growth and ability to receive and deliver the water, and they guarantee that the state will deliver the full entitlements.

Since the project currently delivers only about 2 million acre-feet a year, no one was surprised in the mid-1970s when the state announced its plan to construct Phase Two of the State Water Project. The plan's centerpiece would be a 42-mile canal around the periphery of the Sacramento-San Joaquin Delta to improve water quality within the Delta and to correct some serious environmental problems there, as well as to increase capacity in the State Water Project so that the state's contracts could be fulfilled. Leading the campaign for the Peripheral Canal was the Metropolitan District of Southern California along with its traditional ally, Central Valley agribusiness.

Northern and Southern Californians have haggled over water for decades. But not since the 1850s have citizens in the North and the South gone at each other's throats so viciously as they did during the long debate over the Peripheral Canal.

In 1977, California Senator Ruben Ayala, a San Bernardino Democrat who chairs the Agriculture and Water Resources Committee, introduced legislation to implement the state's plan for completing the State Water Project. His bill focussed on the Peripheral Canal. This unlined, 400-foot-wide ditch would take water from the Sacramento River and carry it around the eastern edge of the Delta to the state's giant pumps at Tracy. It would replace the state's current practice of sending water from the Oroville Dam and the Central Valley Project's Shasta Dam through the Delta's 1000 miles of meandering channels. The canal was also designed to eliminate the reverse flows that suck seawater from the San Francisco Bay into the Delta; thus, it would improve water quality for both local and Southern California use. Most importantly, it would allow the State Water Project and the Central Valley Project to draw an additional million acre-feet of water every year without damming any rivers —politically a big plus.

Governor Edmund G. Brown, Jr.—"Jerry" Brown, son of "Pat" Brown of SWP fame—up for reelection in 1978, endorsed the plan, which seemed to have something for everybody. Best of all, according to *New West*, "the Peripheral Canal would have been a lovely gift to the wealthy agricultural interests who have never forgiven Brown for his chummy relationship with Cesar Chavez."[30]

Just as his father had done 20 years earlier, Jerry Brown assembled the state's diverse water interests and attempted to construct a coalition to support the Peripheral Canal package in the senate. To mollify Northern California concerns, he and his backers in the legislature included in the plan environmental guarantees to protect the Delta and San Francisco Bay; they also included greater statewide conservation measures. His master plan collapsed when powerful environmental groups, including the Sierra Club and the Environmental Defense Fund, rejected it. EDF's Zach Willey called the Peripheral Canal "a loaded gun pointed at the head of the Eel River," because its immense capacity

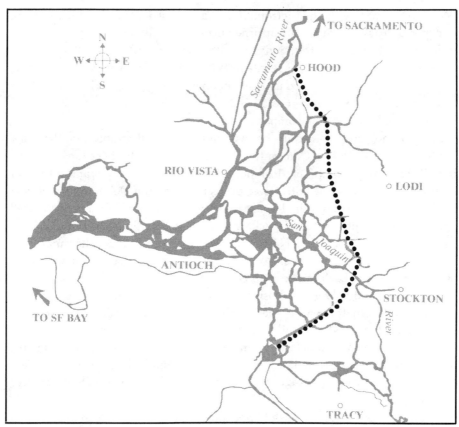

Sacramento Delta and proposed Peripheral Canal from *Water & Power,* by Harry Dennis

would have inevitably led to development of yet another Northern California water supply.

Governor Brown had better luck with the state's biggest water users, Central Valley farmers and the MWD. They were pleased to support plans for more water flowing from the North to the valley and father south, so pleased they donated their considerable resources for lobbying for the canal in the legislature and later in the media.

Brown was unable to convince farmers in the Delta that the canal did not threaten their livelihood. Delta farmers irrigate their land by pumping water directly out of the natural channels; they feared that diverting more water through a Peripheral Canal would leave their land dry and choked with salt.

"We don't want to give the Metropolitan Water District the plumbing to suck us dry in a drought year," Delta attorney Dante Nomellini stated.[31]

Despite lobbying by the environmental groups and the Delta farmers, Senator Ayala's bill passed the senate in 1979 and picked up a few amendments in the assembly. It also picked up new opponents, two giant agribusiness firms in the San Joaquin Valley—the 65,000-acre Salyer Land Company and the 155,000-acre J. G. Boswell Company.

Salyer and Boswell joined the opposition to the canal not because they saw eye to eye with the Sierra Club or the Delta farmers, but because they believed the bill had far *too many* environmental stipulations. They wanted to wait until a "better bill"—one with fewer protections for the Delta and the Bay—could be pushed through the legislature. Another reason Salyer and Boswell opposed the Peripheral Canal package was their fear that MWD would not support future efforts to dam Northern California rivers if it got the canal.

For the first time in 20 years, MWD and agribusiness were on opposite sides of a water debate. Each armed with enormous PR budgets and well-paid lobbyists, MWD and Salyer-and-Boswell turned their attention toward a Southern California senator whose vote on the canal bill was crucial. In 1979 the agribusiness giants won round one; Senator William Campbell, Republican floor leader from Hacienda Heights and a gubernatorial hopeful, voted against the bill, killing it that session. But in early 1980, Campbell switched sides. Senate Bill 200, the canal package, was approved in the same senate committee that had killed it the year before.

"What happened was dozens of appearances by Met officials in and around his district," *New West* revealed in 1980. "They showed their movie [*The Missing Link*, a $40,000 pro-canal film] and warned of a water-short Southern California and mentioned Campbell as the stumbling block." A lobbyist from the anti-canal camp stated, "They really did a number on him."[32]

Its principal opponent now a reliable ally, SB 200 passed the legislature in June 1980. Like Burns-Porter 20 years before, the vote was strikingly regional. Every senator from the Central Valley and Southern California voted for it, and every senator from Northern California voted against it.

Strange Bedfellows

Even before SB 200 passed, environmentalists were convinced that the canal would make it easier, even necessary, for the state to dam the Eel River. They responded by qualifying a ballot measure to require constitutional guarantees against damming the North Coast rivers and to provide stricter guarantees for Delta and San Francisco Bay water quality. California voters passed Proposition 8 in November 1980, tying approval of the Peripheral Canal to these environmental restrictions.

On July 18, 1980, Governor Brown signed SB 200 into law. That day, a coalition of environmentalists—hardliners still not satisfied with the bill's "illusory guarantees"—and consumer advocates and Delta farmers, launched a signature-gathering drive to place a referendum to overturn SB 200 on the next statewide ballot. This would give the state's voters the final say on the Peripheral Canal and its related facilities. Within 90 days, the coalition had gathered 850,000 signatures, more than twice the number needed to put the measure on the ballot. Proposition 9, the Peripheral Canal referendum, would go before California voters on June 8, 1982.

By this time it was clear that just about every special interest in California had something to win or lose if the canal were built. The campaigns for and against Prop. 9 brought them all out and inspired some of the most bizarre political alliances the state has ever seen. Beyond the obvious North vs. South split, the state's farmers were divided among themselves. Delta farmers had been joined in their opposition to the canal not only by Salyer and Boswell but by the influential California Farm Bureau Federation and its 100,000 members. These growers, a number of them corporate subsidiaries of huge land or energy companies, allied themselves with environmental groups like Friends of the Earth and the Sierra Club, usually their arch enemies. Thus, the "strange bedfellows coalition" was born.

Central Valley growers, as mentioned earlier, opposed the canal legislation because of its too-stringent environmental protections for Northern California watersheds. The environmentalists felt exactly the opposite; their feelings about the environmental guarantees of SB 200 and Proposition 8 are summed up in former State Senator Peter Behr's memorable phrase: "a thirsty beast cannot be contained in a paper cage." Along with utility rate reformers and several Northern state legislators—Tom Bates of Oakland and John Garamendi of Walnut Grove, among others—the environmentalists linked opposition to the canal with demands for an overhaul of the state's outdated water management policies and water rights laws.

The strange bedfellows focussed on the outrageous costs of the canal package. Throughout the campaign they countered every official estimate with one of their own, which was often several *billion* dollars higher. They said the state had forgotten to add in skyrocketing energy costs or account for inflation and unproven technology. Their high number for Proposition 9: $23 billion.

A reluctant Jerry Brown, Southern California politicians, agribusiness, labor, and land developers line up to support the Peripheral Canal

Both canal proponents and opponents spent unprecedented amounts of money to win voters to their side. Canal advocates spent an estimated $2.7 million to buy media time and advance their case; Getty, Shell, and Union Oil, all big landowners in San Joaquin Valley who would receive water from the canal, chipped in about $600,000. Opponents spent $3.6 million trying to convince voters "It's just too expensive"; Salyer and Boswell, the only well-heeled members of this group, contributed the bulk of that, about $2 million.

In addition to doing effective door-to-door canvassing, anti-canal forces bought television and radio time and aired slick ads portraying gangsters trying to steal voters' money.

On the pro side, the MWD led the efforts to persuade California voters to pass Proposition 9 so that Southern California wouldn't go thirsty.

"If you have a sustained drought you'll have a sustained change of life-style," stated Earle C. Blais, chairman of the MWD board. "Acreage will be taken out of production, companies will not have sufficient water to meet their needs, which will result in unemployment."[33]

The Supreme Court-mandated reduction in the district's allotment of Colorado River water and other possible losses figured prominently in the pro-canal arguments. Met officials insisted these losses would force them to rely much more heavily on state-project water beginning in 1985, when the Central Arizona Project was set to come on line. (After a protracted legal battle between California and Arizona over each state's legal share, the US Supreme Court in 1963 ordered California to reduce its take. Unsettled suits over Native American claims could mean less water for the Golden State.) California will

An unlikely coalition opposes the Peripheral Canal: environmentalists, Central Valley agri-giants, outdoorsmen, and Mike Curb, the Lieutenant Governor.   Mary Robertson

lose about 20 percent of its current deliveries, most of which go to the MWD and the Imperial Irrigation District, beginning in 1985.

Met officials also argued that current SWP supplies would be severely strained by increasing demands in Northern California where upstream users plan to divert more water and several water districts which have never taken their share of state water will begin to do so in 1990. Not least of all, they pointed out, the district and the other SWP contractors have legally binding contracts to take the full 4.2 million acre-feet of water by 1990. If the canal or an equivalent facility were not built, MWD warned, the state would be liable for breach of contract.

Backed by MWD's massive public relations muscle, the district showed films and sent speakers to talk to every group that would let it. And MWD broadcast slick media spots predicting Southland residents and industries running out of water within a few years if the canal weren't built. Pro-canal television spots were filmed in the Arizona desert.

For the first time in history, many Southern Californians didn't believe the MWD's predictions of a parched doom. According to Mervin Field, director of the California Poll, Southland residents showed, as election day neared, increasing sympathy with canal opponents' charges that the price of Prop. 9 was simply too high. By June 8, it was clear that MWD had failed, in spite of its $1.8 million annual PR budget.

On that day California voters rejected the Peripheral Canal and other facilities proposed in Prop. 9 by an astounding margin—62 percent to 38 percent. In the Southland, 61 percent of the voters did vote for Prop. 9 while 39 percent voted against it. Because Southern California has more voters than does Northern California, one might have assumed the South's greater numbers would have carried the day. But fewer voters in the Southland made it to the polls, and they weren't nearly as united in their attitude toward the canal as Northern Californians were.

Voters north of the Tehachapis were nearly unanimous on Prop. 9, another scheme to send more of their water south. A full 91.9 percent of the voters in the 48 northern counties yelled "NO WAY!" to the Peripheral Canal and clinched its demise.

Ninety-seven percent of Marin County's voters—still stinging from memories of Mission Viejo's swimming pools filled to the brim during the great drought of 1976–1977, while they showered with a bucket at their feet so they could flush their toilet with the "graywater" once a day—voted against Prop. 9. In Contra Costa County, 96 percent of the voters opposed the canal; in San Francisco and Alameda counties, 95 percent voted against Prop. 9; 93 percent in San Mateo County and 89 percent of the voters in Santa Clara County, which receives State Water Project water, punched "No" on their ballots.

"Margins like this are common only in totalitarian countries like North Korea," the *San Francisco Chronicle* noted.[34]

"I have followed politics for 30 years and I've never heard of anything like that," stated David Sears, a University of California at Los Angeles political science and psychology professor.

Many theories have been advanced to explain the amazing June 8, 1982, defeat of the Peripheral Canal package. The most nitty-gritty one, from pollster Mervin Field, says the key issue was the huge pricetag of Prop. 9. In exit polls that day, 53 percent of Northern Californians who voted against it cited the proposition's high costs as their main reason, while 67 percent of Southern Californians who voted "No" said Prop. 9 would cost too much.

Another factor that probably aided the anti-canal side is the demonstrated tendency of voters to vote *against* a proposition they aren't sure about. The Field Institute confirms that when the public is confused about a measure or when there's little or no publicity on one (unlike this case), voters usually vote against it.

Carl Pope, executive director of the League of Conservation Voters, which ran the canvassing operation against the canal, said his side spent a lot of money to educate Southern California voters "that the water issue is very complicated and that dams and canals alone are not necessarily the solution." He said his side's campaign probably made about half of the Southland's voters "realize that there were two sides to the question. There have never been two sides to water issues in the south. MWD has always been predominant."

Southern California voters, no doubt, lost faith in their monolithic water agency. Lawrence Kapiloff, the San Diego assemblyman who sponsored SB 200, said, "The Metropolitan's past finally caught up with them—the *Chinatown* syndrome,"[35] referring to the 1974 Academy Award-winning film based in part on the Owens Valley scandal. The Owens Valley, it seems, has come to symbolize the greed and corruption of water agencies in general.

Neanderthals and Scapegoats

Regional hostility between the North and the South played a major role in the campaigns for and against Prop. 9, whether canal proponents or opponents wanted it to or not. Age-old stereotypes and rivalries die hard.

Usually apathetic about such matters, Southern Californians took a mild interest in the Peripheral Canal debate. Many felt that the North was simply being greedy with its abundant supplies. A typical southern view was expressed by a Malibu hotel developer the month before the election: "There's enough water in Northern California, right? And Southern California needs the water, right? So what's the problem? I don't know why the north resents the south taking their water. It's only drainage to them and lifeblood to us."[36]

MWD public information officer Howard Williams likened Northern California opposition to the canal to "a Neanderthal suspicion—don't go near my waterhole."

Northerners felt a similar disdain for the South's motives. Some called the canal a "demon ditch," designed to suck northern rivers dry for the benefit of an insatiable beast—Los Angeles and the urban sprawl of the Southland.

Spencer Michels, producer of a 1981 KQED documentary called "Two Californias," insisted that the North-South rivalry was *the* most influential factor in the debate. "Even if they understood all the complex political, environmental and economic issues involved in the canal debate," he wrote in *California Living*, "northerners would still feel they were being cheated by a peripheral canal that [would] divert up to 80 percent of the flow of the Sacramento River" and ship it south. "What was the north getting in return? Kern County vegetables?" he asked rhetorically.[37]

Field pollsters discovered a strong regional split of pro and con sentiment throughout the debate on the canal. (See chart next page.)

The North-South hostilities most certainly affected the Prop. 9 vote—exactly how much cannot be quantified, but speculation on why the

North was so united in its opposition runs the gamut from the philosophical and the anthropological, to historical trends and psychological motives.

Carl Pope says the North's nearly unanimous vote on the canal reflects Northern Californians' "sense of cultural cohesion and identity," which simply doesn't exist in the South.

Robert Sommer, a psychologist at the University of California at Davis, views the Prop. 9 vote as "a south vs. north issue. Any good northerner should vote to save 'our' rivers. 'They' wanted 'our' water." He said people in Northern California "identified with a setting that has rivers, mountains, trees, and these people were trying to steal it, that was the perception."[38]

David Sears, another psychologist, but from south of the Tehachapis, sees the North's unanimity differently: "It does tap into something that is not psycho-

The evolution of canal sentiment in the North and the South.

SOUTHERN CALIFORNIA	Favor canal (YES on 9)	Oppose canal (NO on 9)	Undecided
June 1982	45%	27	28
May 1982	52%	23	25
March 1982	68%	15	17
January 1982	57%	16	27
October 1981	59%	15	26
August 1981	55%	13	32
April 1981	53%	15	32

NORTHERN CALIFORNIA	Favor canal (YES on 9)	Oppose canal (NO on 9)	Undecided
June 1982	10%	80	10
May 1982	10%	77	13
March 1982	16%	71	13
January 1982	23%	64	13
October 1981	27%	62	11
August 1981	22%	62	16
April 1981	21%	63	16

Source: The California Poll

logically very different from powerful nationalism or ethnocentrism, including things like stereotyping, scapegoating, and displacing aggressions to a remote place." He said Southern California suffered because of its stereotype as a place where people waste natural resources—"city slickers who would ravage the countryside. There's a feeling in Northern California of Southern California being artificial, that we're a desert that's artificially sustained. . . . and somehow Northern Californians are real."[39]

It is impossible to say how much truth lies in what the psychologists and political analysts say about the June 8, 1982, vote. We may never know to what extent Northern and Southern California's deeply rooted antagonisms and cultural rivalry affected the Peripheral Canal's defeat. One thing is certain. The demise of Proposition 9 in 1982 did not end California's water wars, nor did it close the book on further water development in the state.

What's Wrong With a Little Dam, Anyway?

Why are so many Northern Californians and environmentalists dead set against water projects? California has prospered immeasurably from engineering feats that have stilled rivers, inundated canyons, and piped water all over the state. Aren't they proud of the fact that, other than the Great Wall of China, California's monumental dams and aqueducts are the only man-made structures on Earth visible from the moon? A little water project doesn't hurt anybody, does it?

A *little* water project may not hurt too much, but California's water projects are generally not "little." Opponents of more water development are not simply nay-sayers; nor are they Neo-Luddites who are afraid of progress. California's modern water system may be responsible for saving Central Valley farmers from ruin 30 years ago and for enabling the Southland to boom and bloom. But the cost of all that concrete to the state's environment has been dear, and the time when large water projects contributed to the welfare of society has long since passed. The best sites have already been developed, experts agree, and the pricetags on big water projects have made them unpopular both with politicians and the public.

Every place a dam, canal, aqueduct, or channel has been built, the consequences of human tampering with nature are evident. Most Northern California rivers, once rushing whitewater tumbling down wild canyons, have been reduced to placid streams or tranquil reservoirs. Besides eliminating whitewater rafting and sport fishing, "with dams and diversion structures, we have smoothed out the seasonal peaks of natural streamflows and altered the concentration of sediments and nutrients these rivers once carried," states the *California Water Atlas*.[40]

"As a result, the modern system of dams, reservoirs, and artificial channels has encouraged erosion in some areas and stopped it in others, while slowing or halting the formation of alluvial floodplains in some parts of the state and accelerating their formation in others."[41]

Moving copious amounts of water around the state has also changed the distribution and abundance of every native aquatic plant and animal in California. Hardest hit by the changing waterscape have been native waterfowl and shorebirds.

"No better barometer" for determining the health of California's river ecosystems exists than the yellow-billed cuckoo, says the *Atlas*. This songbird originally nested in willow and cottonwood trees in most coastal valleys from San Diego to Sebastopol. It also flourished throughout the Central Valley from Bakersfield to Redding, in the Owens Valley, and along the Colorado River. But the cuckoo's breeding grounds disappeared as groundwater levels fell from irrigation pumping, as streamside vegetation was cleared for flood control, and as marshlands were "reclaimed" for agriculture and forests cut for wood. In 1977, only 35 to 68 pairs were reported in the Sacramento Valley; the species is now considered a rare bird in California.[42]

In place of the yellow-billed cuckoo and other riparian songbirds has come the cowbird, a parasite that deposits its eggs in other birds' nests to be hatched and fed. Before 1900, only one cowbird was seen in the Central Valley. Now flocks of up to 10,000 are common along the Sacramento River.

The salmon is another good index for assessing the environmental impacts of our modern waterworks. Once a rich fishery in all foothill tributaries of the Sacramento and San Joaquin Rivers and in the larger coastal streams, the king salmon has plummeted as a result of increased sedimentation in its previously clear streams.

What has happened to the Trinity River in the rugged far north illustrates how water development affects a river's fishery. The Trinity has been dammed for decades, and its fishery has crashed. State and federal hatchery efforts, writes Harry Dennis in *Water & Power: The Peripheral Canal and Its Alternatives*, have had little effect on the problem. Silt from logging and road building used to be flushed out by the river's heavy winter flows, but now "80 to 90 percent of the Trinity's annual flow is diverted through a tunnel into the Central Valley. The river can no longer scour away the silt from its once-gravelly spawning beds. Fish eggs, if laid, cannot survive in the silt."[43]

A study conducted for the Bureau of Reclamation and published in 1979 found that king salmon and steelhead trout have suffered the most from the Trinity River diversions. It states: "The number of adult king salmon returning to the hatchery during the post-1969 period is about 50 percent of that estimated to have passed above the side of Lewiston Dam in 1944 and 1945 and about 15 percent of that estimated in 1955 and 1956. . . . The number [of steelhead trout] spawning in the tributaries above the North Fork declined 90 percent from 1964 to 1972."[44]

In addition to siltation, anything that acts as a barrier can reduce fish populations and interfere with spawning—including higher water temperatures and the concentration of pollutants in a stream.

Reduced flows from water diversion and dams can delay the start of salmon migration, and many thousands of young fish are killed every year in the huge hydroelectric turbines, by the poor water quality in reservoirs, and by predators who wait for them to collect along a dam's walls.

The Delta

Nowhere have the environmental effects of water development been felt more acutely than in the Sacramento-San Joaquin Delta, the centerpiece of both the state and federal projects. The Delta is the largest inland river estuary in the continental United States. It consists of 1100 miles of channels and hundreds of thousands of acres of islands through which move California's two most hydrologically important rivers—the Sacramento and San Joaquin.

Near Tracy at the southwest corner of the Delta, two giant pumping plants, one belonging to the state and one to the federal government, draw water for distribution elsewhere in California. The pumps are "two straws in the same glass," say Tom DeVries and George Baker. "The institutions that draw on the straws are hearing the nasty noises of a nearly empty portion and they glare at each other and quarrel as they suck."[45]

Reverse flows are the problem here, the intrusion of sea water from San Francisco Bay into the Delta's freshwater environment. The 1977 drought was a harbinger of what could result from increased freshwater exportation from the Delta. Populations of shrimp, striped bass, and other wildlife declined dramatically. Many ecologists are convinced that increased freshwater export from the Delta could destroy or seriously damage the whole seagoing food chain in San Francisco Bay. Nature abhors a vacuum. If fresh river water is removed, nature will replace it through tidal action with salt water. The result: a drastic ecological change. Eventually, crab, shrimp, oysters, salmon, bass, and other delectable and aesthetically valuable wildlife would disappear. National Marine Fisheries Service biologist Susan Smith can think of only one creature that would proliferate in the new, more saline environment of the Delta—sharks.

How do fish in the Delta now fare? Not too well. "Striped bass and eggs are currently ingested by the uncounted millions in the existing pumping plants [at Tracy], which also interfere with salmon migration," Baker and DeVries report.[46] Although it has tried, the state has not been able to devise a fish screen that will keep fish eggs and very small fish from passing through. If the young do slip through the screens at the intake pumps, they either die along the way, end up stranded on fields in the Central Valley and then die, or make their way into aqueducts and reservoirs in Southern California where they soon expire.

San Francisco Bay

In 1981 Floyd Ander, acting director of the National Marine Fisheries Service, wrote that the declining populations of fish using the Delta and Bay estuary are "very likely the indirect result of export of fresh water from the estuary. The exporting of water reduces Delta Outflow [the amount of water that flows through the Delta and out into San Francisco Bay], which has been shown to control the primary productivity and water quality in the [Bay]."[47]

Adult bass counts in San Francisco Bay are way down—about one-third of the high counts taken during the 1960s. The NMFS has discovered that half of those suffer from lesions, parasites, and deformities, no doubt the result of increasing pollution in the Bay.

The drop in fish populations may just be an early warning signal for bigger environmental problems. Dr. Michael Rozengurt, who studied the effects of diversions from delta ecosystems similar to the Sacramento-San Joaquin Delta but in the Soviet Union, asserts that fresh water in sufficient amounts is needed to flush pollutants and salty water out of a delta. Rozengurt believes that further reducing the amount of water flowing into such an estuary could be devastating. Currents that now take sewage out of San Francisco Bay, for example, could reverse and "bring it right back in." From his studies in the Soviet Union, he concluded: "No more than 25 to 30 percent of the natural outflow of an estuary can be diverted without disastrous ecological consequences ensuing."[48]

Among the early warning signals he observed were "reduced productivity of fish and wildlife resources, changed biological structure of plankton [which has been also observed in San Francisco Bay]. . .increased salinity intrusion affecting municipal and agricultural water supplies and the biota, and increased effects of pollution loads in progressively stagnant waters."[49]

Before the Gold Rush, an average of 30 million acre-feet of river water flowed from the Delta into San Francisco Bay every year. Today only about 13.5 million acre-feet make the trip. State water planners expect to cut the flow to well below 11 million acre-feet by the year 2000. In wet years, plenty of fresh water would still reach the Bay, but in dry years, about three in ten, water agencies would probably allow only 3.6 million acre-feet to flush the Delta, and less than that in critically dry years. If Dr. Rozengurt is correct in his predictions, such planning could wreak havoc on the entire San Francisco Bay-Delta Estuary.

The Valley

Changes in the natural landscape from water development for irrigation and the Third California have been dramatic in the Central Valley. Having spent a hundred years pumping up groundwater and piping rivers and streams across whole regions, farmers now feel an ill wind blowing.

Formerly a vast expanse of alkali flats, grassland prairie, and marshlands, the Central Valley is now blanketed with bright green fields growing all sorts of food and fiber crops. Like everything else, such abundance has a price. As we mentioned earlier, the valley's farmers had to be bailed out in the 1930s when they depleted their groundwater basins. The Central Valley Project and later the State Water Project were intended to put a stop to the mining of groundwater. But they did no such thing.

The Institute of Rural Studies has reported that state project water has generally been used to bring new ground under cultivation—not to relieve the overdraft on lands that are already irrigated. At least 250,000 previously unfarmed acres have been planted as a result of the availability of cheap water in the valley, particularly along the west side of Kern County.

"One now sees irrigated fields creeping over the bleak Coast Range foothills," writes Marc Reisner, "and spreading into hellish desert valleys around Bakersfield."[50] In Kern County alone, 1,340,181 acre-feet of state water was used in 1981, according to Gary Bucher of the Kern County Water Agency. But only 234,034 acre-feet went directly to recharge the aquifers in the county. The rest went for irrigation (more than a million acre-feet) and household and industrial use.

Many of the new fields are planted in heavy water-using crops, such as rice, cotton, pasture, deciduous orchards, tomatoes, alfalfa, and sugar beets. The low price of state and federal water actually encourages the cultivation of these water-intensive crops as well as wasteful water practices.

"Rice, alfalfa and pasture...together with other crops grown for livestock use—corn, milo, grain—account for 40 to 50 percent of the state's total irrigated acreage," the *California Water Atlas* states.[51] These crops don't have a high enough value, in general, to pay the cost of the water they require, which simply perpetuates the need for continued subsidies and exacerbates the groundwater mining. Overdrafting in the valley continues at 1.5 to 2 million acre-feet a year.

Because the valley supports more agriculture than God ever intended it to, and because irrigation water is often salty and modern farmers rely heavily on chemicals, the valley's soils are becoming increasingly saline. Much of the valley has an impervious soil layer below its fertile topsoils. So the mineral salts left over from fertilizers and pesticides and the brackish water used to irrigate get trapped. Eventually they make their way back up to the topsoils. "To maintain productivity, farmers must continually wash salts out of the root zone of their crops," Dennis states. "This is done by giving the crops more water than will be taken up by the roots, so the excess water will percolate below the root zone, carrying the salts down with it."[52]

In the west side of the valley, drainage is slow. After a few years, the brackish water collects and reaches the root zone of the plants.

When this happens, flushing is useless. In some areas, the soils have become so salty that farmers have had to switch from fruits and vegetables to salt-resistant crops, such as cotton, corn, and other grains. Eventually, the soil will become too salty for any crops.

So far the problem affects about 400,000 acres. Without proper drainage, growers and soil experts alike predict that over a million acres will go out of production—the land reverting back to its original alkalinity—by the year 2000. San Joaquin Valley farmers want a $750 million drainage system built to take the salty water away and dump it—you guessed it—in the Delta.

Groundwater overdrafting has also led to land subsidence. Since the 1920s, the land surface in western Fresno County, for example, has gone down 29.6 feet from intensive pumping of well water. The land is sinking in about 10,000 square miles of the valley. A recent US Geological Survey study concluded: "Subsidence in the San Joaquin Valley probably represents history's greatest single man-made alteration in the configuration of the Earth's surface."[53]

Farther south, in the scorching Imperial Valley east of San Diego, farmers receive so much irrigation water from the Colorado River, they actually waste about 400,000 acre-feet a year, according to a recent state study. But Colorado River water is getting saltier as more and more upstream farmers use it and then dump the wastewater back into the river. By the time it reaches the Imperial Valley, it's full of mineral salts, so the farmers must also flush their croplands with extra water. Growers in the Imperial Valley have one advantage over farmers in the slowly draining west side of the San Joaquin Valley. They have a place readily available to dump their wastewater. They let it drain into the Salton Sea, an accidentally man-made lake—but, of course, the sea is filling up and beginning to overflow. And the concentration of pollutants and salts is threatening to wipe out the birds and fish that settled there after 1905 floods overflowed an early irrigation canal, filling up the Salton Sink and turning it into California's largest inland sea.

Mono Lake

Of all the harm human beings have inflicted on California's waterscape in the name of "water for the people," the shrinking of Mono Lake stands out as the most visible and surely the most tragic. Since Los Angeles began taking fresh water from Mono Basin feeder streams, the level of the lake has dropped 46 feet; today a broad, white bathtub ring can be seen around Mono's shores.

Mono Lake, a strangely beautiful place that Mark Twain called "the loneliest spot on Earth," is a saline remnant of a vast inland sea that covered about 316 square miles in the Mono Basin and nearby Aurora Valley more than 13,000 years ago. Early explorers nicknamed it the "Dead Sea of the West" because of its eerie-looking tufa towers and salty water. The lake actually supports myriads of microscopic organisms, flies, and brine shrimp. Mono Lake attracts thousands of gulls, grebes, and other migratory birds who stop there to breed and rest on their way south. Until 1979, Negit Island in the middle of the lake supported the second largest rookery for California gulls in the world.

When the Los Angeles Department of Water and Power's Mono extension went into operation in 1941, the level of the lake started dropping a foot a year. Since the second aqueduct began diverting more water from the basin, the lake has been shrinking about 1.6 feet per year. It is not known if the brine shrimp, the gulls' chief source of food, can survive the increasing salinity that is coming as the lake's volume decreases. Declining water levels have threatened the gull rookery on the lake's islands as well, by creating land

Mono Lake, 1962, courtesy Eban McMillian

bridges, which enable mainland predators to cross onto the islands and eat eggs in the gulls' nests.

If you drive over the Sierra to Mono Lake, you come to the town of Lee Vining, population 300. It's a typical small town, with assorted gas stations, motels, and coffee shops, but the people who live there are witnessing one of the most dramatic examples of what can happen to a natural body of water when someone with more money and more power wants it.

"It is as if you can see the lake dying," says Lily Mathieu, a descendant of pioneers in the area.

The Mono Lake Committee makes its headquarters at Lee Vining. "We don't want to restore Mono Lake to its pristine condition," says David Gaines, chair of the committee. "We only want a compromise: enough water to keep the lake alive." He claims this is possible if the people of Los Angeles put a brick in each of their toilets. But it is the Department of Water and Power that holds exclusive rights to Mono Lake water, not the citizens of Los Angeles. And the Department "won't give up a single drop," Gaines sighs.[54]

In spite of the lawsuits demanding that it do so, the LADWP has shown virtually no willingness to decrease its diversions from the Mono Basin. It derives 20 percent of its water supply from the basin and makes a pretty penny selling the hydroelectric power its plants generate along the downhill, southward journey. William Kahrl, author of *Water and Power, The Conflict over Los Angeles' Water Supply in the Owens Valley*, insists that the city could

Mono Lake, 1978, courtesy the Mono Lake Committee

get along just fine without taking so much water from Mono Basin: "The years when Los Angeles required a super-abundance of water to serve an ever-expanding population have passed; the city's population has in fact stabilized since 1970." He says the LADWP still takes so much water from Mono's feeder streams "to enable [it] to make use of the water rights it had already acquired and to assure that it would be able to continue providing its customers with the highest quality water available at the lowest possible price."[55]

The city could, of course, get much more of its water from the Metropolitan Water District of Southern California, which has a huge surplus each year, but Los Angeles prefers the cheap water from Mono. By the mid-1970s, the city had reduced its MWD supplies, which cost twice as much as its aqueduct water, to less than 3 percent of its overall supply.

Conservationists are suing the Department in an effort to reduce the devastating diversions. They are attempting an unprecedented legal strategy: to declare the DWP's diversions a violation of the public trust.

Sherman's March to the Sea

Owens Valley, Mono Lake, Hetch Hetchy, the Trinity, the Stanislaus, what will be the price of the next round of water development in California?

A couple of years ago, Greg diGiere, administrative assistant to California State Senator Barry Keene, predicted, "If the Metropolitan Water District and the Central Valley farmers get together and agree on what they want, there won't be a whole lot that the rest of the state can do to stop them. It will be like Sherman's march to the sea."[56]

Well, the old allies are back together again. Without a moment's hesitation, San Joaquin Valley growers opposed Proposition 13—"The Water Resources Conservation and Efficiency Act"—on the November 1982 ballot. The measure would have radically changed the way water is used in California and how much water users pay for it. MWD came out against Prop. 13, too, and it was slaughtered at the polls. (See chapter 6 for more on Prop. 13.)

If the big water powers in the state do get their way, no doubt they'll dam the Eel River and turn the last wild river in California into a placid reservoir (they might even have the chutzpah to name the project after William Mulholland or Fred Eaton). If the new governor sides with the concrete worshippers—as early evidence indicates he will—and he convinces the legislature to go along, Central Valley growers will get a through-Delta channel so more Northern California water can be exported south, and they'll get their agricultural drain to dump polluted water into the Delta. The net effect of such a scenario just might be to turn the fragile Delta estuary into an enormous salt flat and the San Francisco Bay into a stinking sump incapable of supporting any species of life, save sharks.

If this does happen and Los Angeles refuses to reduce its Mono Basin diversions and the lake shrivels up, its shrimp die, and its seagulls quit the scene for good, ecologically minded Californians would be angry. Northern Californians, for sure, would be livid; if they could muster up the kind of unanimous opposition to such depradations on their watersheds that they exhibited in the June 1982 vote against the Peripheral Canal, it could be enough to trigger a serious drive to split the state. Another big porkbarrel water project might convince environmentalists around the state, frustrated in their attempts to establish critical water conservation measures and pricing reforms, to change their minds about the wisdom of a North-South divorce. With the environmentalists' powerful support, a split-state movement could very well succeed. If it did, the unquenchable thirst of the Third California would be as much to blame for ripping California in two as would the age-old competition between Northerners and Southerners.

5

Imagine Two Californias

ruthless California governor wants the states west of the Rocky Mountains to secede from the Union and become an independent nation. He will stop at nothing to see his dream of the great "Western States of America" come true. Once the new country has gained independence he will be its first president —or emperor. Faced with a crippling Islamic oil embargo, a runaway constitutional convention, and the power-crazed governor, the United States comes dangerously close to severing itself into two separate and not necessarily friendly countries.

Such is the plot of the exciting 1982 novel *The Great Divide* by Frank Robinson and John Levin. Although the writers have clearly let their imaginations run wild, the book taps into some profoundly disruptive trends in the nation today. Separatist movements—motivated by everything from taxation without representation to desire for cultural identity—are proliferating across the land. The people of south Jersey, for example, want nothing to do with the cosmopolitan Manhattanites in northern New Jersey. People in Alaska and the Dakotas want out of the

> **"Politics is the only forum in which we can resolve our differences. As long as equally worthy people have incompatible goals, somebody has to mediate. Unless you want things decided by the whim of a dictator or unless you want to shoot it out, the politicians are our mediators."**
>
> **—John Gardner**

159

oppressive federal government, as do the angry fishermen of Florida.

The Sagebrush Rebellion a few years back was just one incarnation of the West's hostility toward the federal government. The taxpayers' revolt that brought us Proposition 13 in California, Proposition 2½ in Massachusetts, and a host of tax-slashing measures around the country attests to Americans' dissatisfaction with expensive, inefficient, centralized governments. Federalism, it seems, is suffocating under the weight of its hefty bureaucracies. The citizens of America want more local control, better representation, a government responsive to their particular needs. All across the land you can hear people shouting "Give the government back to the people!"

In California, it is Northern Californians who complain the loudest about lack of control over their lives and resources. Since 1965, when the Supreme Court decided that legislative representation would be based on the one-person, one-vote principle, the North has been held captive by the more populous Southland. Every so often, a group of Northerners will sound the battle cry, "Split the State!", demanding their rights as equal Californians.

Right now the movement to divide the state simmers on a back burner. But as we have seen, it is always ready to boil when the legislature does something particularly jarring to northern sensibilities. Even after Northerners succeeded in killing the Peripheral Canal, resentment against the dominant South persists. In a November 1982 letter to the *San Francisco Chronicle*, an Alameda woman asks "Isn't it about time that the voters of Northern California rebel against having their political leaders chosen for them by Orange County?...If there is any organization out there working to divide this state into North and South, I am ready to join. I realize that at some time maybe a Cranston or Jerry Brown would be sacrificed, but that would be worth it if the Reagans, the Deukmejians, the George Murphys, and the Pete Wilsons were kept in the South."

During our work and research for this book, we were struck by the number of people who, when told the book was about splitting the state, said, "Great! Where's your petition? I'll sign up all my friends," or words to that effect. Northerners still don't like being a satellite of the South, particularly since they feel so superior in culture and intellect. Underlying their snobbishness, however, is the nagging fear that the South won't let sleeping dogs lie; that the powerful water interests in the South will try again for a Peripheral Canal, and this time they won't lose. If that happens, could a split-state movement succeed in California? Let's see.

A Constitutional Right

The first question many people ask about splitting the state is "Is it legal?" The answer is yes. The US Constitution provides for the formation of new states in Article IV, Section 3: "New States may be admitted by the Congress into this Union; but no new State shall be formed or erected within the jurisdiction of any other State; nor any State be formed by the junction of two or more States, or parts of States, without the consent of the Legislatures of the States concerned as well as of the Congress."

This article has been debated numerous times during California's historical split-state drives (see chapters 1 and 2). The lengthiest discussion on the legal issues raised by a potential split may still be a letter Governor Milton Latham wrote in 1860 to President James Buchanan. He sent it along with the California legislature-approved Pico Act to divide the state and the southern counties' favorable vote. Arguing for division, he stated that the Constitution clearly allows new states to be formed within the boundaries of existing states: "The only limitation appears to be in the required consent of the Legislatures of the States concerned."[1]

In 1881, when the subject of division arose again, a committee was set up to examine the legal questions of a split and concluded that the Act of 1859, the Pico Act, was still in effect, despite Congress's failure to pass it; that under Article Four, Section Three, "it only remains for Congress to admit the new state with a republican form of government"; "to secure this last action no legal forms are required"; and that a convention of representatives of all the state's counties where a new constitution would be prepared, approved, and then submitted to Congress would be the best way to accomplish a division of California.[2]

Congress could not be expected to consider the Act of 1859 to be "in full force and effect" today. Nevertheless, a look at how other states were created and admitted to the Union will provide historical perspectives on a possible division of California.

Peter Sheridan, a government analyst at the Library of Congress, prepared a report on the subject in 1980 in reponse to a request by a member of the now-defunct Two Californias Committee. The report, "Procedures for the Creation and Admission of States," reads in part:

"Although no State has been created by combination, creation of States by division has happened in several instances in American history. Five areas had been parts of other States, and were admitted as separate entities by simple acts of admission; these were Kentucky (Virginia); Maine (Massachusetts); Vermont (New York); Tennessee (North Carolina); and West Virginia (Virginia)."[3]

Sheridan points out that West Virginia was formed in 1863 within the jurisdiction of Virginia. The counties in the western part of the state were loyal to the Union and opposed to secession; they set up a "Restored" government in Wheeling and later drafted a state constitution. After the Civil War ended, the people of Virginia approved the constitution as did the US Congress, and the State of West Virginia was born.[4]

When Texas was admitted to the Union in 1845, leaders in the huge territory decided an insurance policy was in order. Wanting a stronger guarantee than the one provided by Article IV, Section 3, they added the following clause to the annexation resolution: "New States, of convenient size, not exceeding four in number, in addition to the said State of Texas, and having sufficient population, may hereafter, by the consent of said State, be formed out of the territory thereof, which shall be entitled to admission under the provisions of the federal constitution."[5]

Although several attempts have been made to divide Texas into as many as five states, none has come close to succeeding.

Sheridan concludes: "It seems clear, then, that a State may be divided or combined, and statehood requested, provided the constitutional requirements described in Article IV, Section 3 are met, plus compliance with whatever standards a particular Congress may require." These have, he says, historically included an exhibition of sympathy for democratic principles as illustrated in the American government, a demonstration that a majority of the people desire statehood, and evidence that the proposed new state has enough population and resources to support a state government as well as its share of the federal government.

It seems reasonable to assume that both Northern and Southern California could come up with a display of sympathy for democracy and that each region would have sufficient population and resources (at least financial) to support its own state government and share of the federal bureaucracy. The real question, then, is whether or not a majority of the people would demonstrate a desire for a new state within California's current borders.

Another legal obstacle to forming a new state exists in Article 10 of California's constitution, which requires that any amendments of its articles be approved by two successive legislatures and then submitted to a vote of the people. Back in 1869, Governor Latham also considered this problem in his letter to President Buchanan. He wrote: "First: Is a change of the boundaries of a State an amendment of the Constitution within the tenth article? And, second: If so, is not that act controlled, so far as a division of the State is concerned, by the clause already quoted from the Constitution of the United States?" Latham does not resolve these questions, stating only that "a change of boundary might, perhaps, be an amendment requiring a vote of the people of the whole state."[7]

Steve Pitcher, a San Francisco attorney formerly with the short-lived 1980 Two Californias Committee, believes that the only way California could be legally split is through the initiative process—a vote of the people, as required by the state constitution.

"It's obvious you can't do it in the legislature, because the legislature is blocked by the Southerners. There's no way they're going to voluntarily split the state. . . . It would be like killing the goose that lays the golden egg. The South is a resourceless area; they don't have our timber, our water, our natural resources. They get too much from us, in the way of taxes and natural resources."

First, you would need a legal petition, he says. Then you must gather the requisite number of signatures to qualify the initiative for the next statewide election ballot. "Now, assuming the people of the state vote to split the state in two, that bypasses the legislature. And the issue goes to Washington, to both houses of Congress. Both the House and the Senate have to vote with the people of California to divide the state, and the President has to sign it," he explains.

"The US Constitution requires that there be a vote of the legislature. So there's a purely legal question as to whether a vote of the people is a valid substitute for a vote of the legislature. I could see this thing going all the way to the Supreme Court—whether a referendum of the people is a valid substitute for the language of the US Constitution. It's never been done by an initiative . . . . It would be a big gamble, to spend a lot of money, go through the initiative process, and get all the way to Washington, and have this thing probably stopped midway between here and Washington and taken to court. As soon as the election was finished here in California, I'm sure someone would say, 'It's all very well and good that the voters have said this, but it's not according to Hoyle.' "

How to Split a State

In 1978, Barry Keene, then an assemblyman from Eureka, introduced a bill to create "Alta California" north of the Tehachapi Mountains. In announcing his bill, he admitted that many technical problems would have to be worked out

Barry Keene, Andrés Pico's inheritor

"to allow the secession of Alta California to take place smoothly." He stated that the new northern state would have to hold a constitutional convention to develop a governmental framework for itself. He added that the southern counties might want to convene their own constitutional convention to revise the governmental framework of their shrunken state. "If southern California wants to call itself 'California del Centro' or just 'California' or anything else, no one up here will object much."

The constitutional convention would have to examine such questions as: Who will pay for the new state bureaucracy and the new state capital? How will each state support the costs of its government? How will the two divvy up the state debt, about $6 billion as of fall 1982? How will the two split up their joint assets—revenues, public lands, property, offices, and personnel? How will various laws and regulations be affected? What will happen to the new set of interstate commerce transactions? Indeed, a division of the state will involve complex organizational, economic, and political questions. The range of questions is limited only by one's imagination.

Addressing the assembly, Barry Keene suggested, "The details of dividing the state bureaucracy would have to be worked out by the two states working together. In some cases, we may decide together to share existing governmental resources, such as data processing ability and operational equipment, for years into the future. We could even decide to share some large departments, for example, one large Department of Motor Vehicles for both states, if we decide together that there are some economies of huge scale that we want to keep."

Keene said three options would be available to work out such questions: the legislature could work them out after the bill passes; the two new states could work them out in negotiations; or the legislature could appoint a joint commission with equal representation from the two new states, as the legislature did in 1859 at the request of the Southern California legislators.

The first problems to be addressed would probably be where to draw the borderline, how to divide the state bureaucracy, and where the new state capital would be. When asked what the Southland would do for a capital, since the North would likely keep theirs in Sacramento, Keene replied, "That's their problem. I suggest Disneyland."[8] (For a discussion of where to draw the borderline, see chapter 6.)

Offices and Employees

As for dividing the state bureaucracy, it is probably safe to assume that the existing legislature could be split into two bodies, with current senators and assemblymen from the North going to the northern state's new legislature and representatives from the South forming the southern state's new legislature. State agencies would be harder to divide because not all have offices in the Southland, and most of their work is done in Sacramento. But with effort, existing bureaucracies could be divided in two, assuming that both states want half of every department and agency. It is possible that South California, for instance, might not care to have a Coastal Commission or that North California might want to dismantle some of the state's drug enforcement agencies. In time, each new state could set up its own bureaucracy and elect a legislature to suit its individual needs.

Barry Keene stated in 1978 that he would favor a unicameral legislature for the North, with districts apportioned by population, as both houses of the legislature are now. He calls the current two-house system wasteful and inefficient and says eliminating it would save taxpayers millions of dollars a year. Keene introduced legislation in 1982 to create a unicameral legislature, but it got nowhere.

Kate Yeager, who prepared an analysis of the pertinent legal, political and economic issues raised by Keene's bill, says that "Local and regional governments will experience changes with the new border: Counties may be split by the change and have to redefine their jurisdictions on both sides of the boundary." She concluded that the current bureaucratic structure would permit the "necessary changes in a cooperative fashion, with very little initial change in policy."[9]

Steve Pitcher seemed optimistic that resolving such matters as the federal ban on interstate branch banking, which would affect several large California banks, and the allocation of federal funds between the two new states would not be terribly difficult. He said he thought a committee would be set up by the legislature to "diddle with all the laws that have peculiar statewide impacts and which would have to be changed. The committee would review all these laws and come up with an interim agreement to accommodate the two states. Most laws are perfectly applicable to two states. Some are not.

"Once the new states are constituted," he added, "all things are possible. None of the issues we considered [in the Two Californias 1980 proposal] are insurmountable. They just require lawyers to be creative, and that's what they're paid for." He smiled.

Booty and IOUs

If California voters and/or the legislature could ever be convinced that Two Californias were better than one, divvying up the state's assets and liabilities would likely be the most troublesome challenge facing the responsible parties —the most troublesome, that is, after the water problem.

As of this writing, California owes approximately $6 billion in outstanding general obligation bonds and is unable to balance its budget. Its bond rating has gone down, making it more expensive for the state to borrow money, and Californians are being warned they could receive IOUs from the Franchise Tax Board instead of cash or checks for their income tax refunds. How two states might resolve this nettlesome set of problems is anyone's guess.

Merely the cost of building a second state capital in the South would be substantial, possibly more than a billion dollars. Santa Rosa biology instructor Frank Gleason, also formerly with the Two Californias Committee, suggests that a deal could be struck between the northern state and the southern one so that "North California could pay for some of the costs of building the new capital, since Sacramento would still have useable buildings for the North's government."

The tale of woe of Alaska's capital, however, points up some very real problems. For years Alaskans complained that the capital in Juneau was inconvenient and hard to get to. Most of the state's people live in Anchorage and Fairbanks and had to take a boat or fly to reach Juneau. So a few years ago, a statewide referendum considered whether to build a new state capital, midway between Anchorage and Fairbanks. It passed, but Alaska is still without its new capital. Money-conscious Alaskans won't pass the necessary bond issues to fund construction. The moral of this story is: a split-state initiative might get through, but it doesn't matter if citizens won't vote for the bond issues to pay for a new government.

The Federal Pie

One of the most troubling fiscal problems facing a newly divided California would be how to apportion the huge amount of money—about $11 billion—California receives from the federal government each year for programs ranging from school lunches and health insurance to timberland preservation and highway construction.

Joe Lang is a consultant to the Assembly Committee on Governmental Organization, for which he prepared a brief analysis of Barry Keene's Alta California bill. Keene's bill would have divided California along the crest of the Tehachapis and east to the Nevada border, so that the southern state would retain approximately 55 percent of the population, leaving the northern state with 45 percent. Lang's 1978 analysis states: "An initial review by committee staff and the Department of Finance indicates that since a portion of federal funding currently received by the state is based on the population of the state as a whole, the division contained in this bill could result in a redistribution of some federal funding with the North receiving less than is currently distributed to it."

He recently elaborated on this: "For example, the North receives most of the federal money that comes into California for fish hatchery programs, which are mostly in the North. California as a general rule receives about 10 percent of all federal funds allocated to the states, based on its having about 10 percent of the US population. If the state were split, North California would have only about 4.5 percent of the US population. So it would lose some of the funding that's allocated according to population, and fisheries in the North could suffer.

"As a whole, we pool our money and get a lot because we are so big," Lang continued. "Even though all federal funding is not based on population, you have to look at the whole question of equity. If the state splits and you have 45 percent of the population in North California and 55 percent in the South, and if the feds didn't change their current levels of funding to each region, South California would be paying an inordinate share of federal income taxes for the amount of benefits it was receiving. Is that fair?"

Northern Californians might counter this with: Is it fair for the North to be paying more taxes in proportion to the benefits it now enjoys from the state government? But who's keeping score?

Jim Harding, former staff member of the California Energy Commission and an economist working at Friends of the Earth, speculates on how the northern state might suffer a loss of federal funding: "Federal monies coming into the two states would be a very contentious subject. If federal grants for flood control, for example, were split down the middle, North California could lose out because the need for flood control is greater in the North."

State budgetary considerations would also have to be made. Revenues flow into Sacramento coffers from all over the state. Revenues from oil production on state lands, primarily in the South, support—among other programs—the University of California system statewide. Harding muses, "Would the UC system in the North suffer if the South kept its oil revenues?"

Dividing up other state assets could be just as tricky. "State parks, for instance—I'm sure some are revenue producers and some are sinks," Harding says. "You might have to liquidate assets to resolve these disputes....It'd be like any other divorce proceeding. The biggest controversy would be over who gets the property held in common, such as public lands, revenue sharing programs, federal grants, etc. South California would say the federal monies should be divided by population, and North California would argue that it should be done by square foot."

How the states would split California's state college system and state and federal funding for highways (60 percent of federal highway funds for California goes directly to the ten southern counties of San Luis Obispo, Santa Barbara, Ventura, Kern, Los Angeles, Orange, San Bernardino, Riverside, Imperial, and San Diego, while the rest makes its way to Northern California), not to mention how they would split hospital, school, and prison funds and facilities,

are further thorny issues to be resolved. Since 51.5 percent of the prisoners currently housed in Northern California facilities—which house two-thirds of all the state's prisoners—actually committed their crimes in the Southland, the two states would have to decide where to put them. In 1977, *New West* magazine suggested building new prisons in Death Valley, "where Charlie Manson has wanted to live all along."[10]

A look at a recent secessionist movement in California will shed some light on how such political and bureaucratic issues might be handled.

The Ponderosa Rebellion: A Model for Secession?

The foothills vs. the flatlands. That's how many people saw the 1982 drive to sever the large County of Fresno in two. Inspired by feelings of impotence and crying "taxation without representation," a group of residents in rural eastern Fresno County decided they'd had enough. They set out to secede and form their own county—to be called "Ponderosa," after an old judicial district. Though their effort failed, it illustrates—in a microcosm—the kinds of difficulties Californians would face in the event of a divorce between the North and the South, and the kinds of solutions we could expect.

The leader of the movement to divide Fresno County was Paul Bartlett, president of KSEE-TV in Fresno. He traces the foothills residents' feelings of abuse and neglect back several years, to when the area was represented by an insensitive, arrogant county supervisor. Bartlett, a jovial but feisty man, ran against the uncooperative supervisor and lost, outvoted by "flatlanders"—residents west of the Friant-Kern Canal.

Then in October 1979, 500 mostly angry people attended a meeting to discuss the foothills residents' discontent with the county government. "That meeting was one of the big surprises of my lifetime," Barlett told Wanda Coyle of the *Fresno Bee* (whose excellent coverage of the Ponderosa movement brought the issues to life). "Not because I didn't believe that the people felt that way, but because mountain people don't turn out en masse."[11]

The Mountain Alliance was formed that night, and 340 ballots were cast in favor of forming a separate county government for the mountain area. By May 1981, the group had gathered enough signatures to qualify a ballot measure for the 1982 statewide election. In fall '81, Governor Jerry Brown appointed five people to the Ponderosa County Formation Review Commission to study the proposal, determine the economic feasibility of the new county and the impact on the remaining county, and prepare a financial and property settlement between the two.

One of the two commissioners from the Ponderosa area, Betty Doyal of Piedra, publicly supported the proposed secession. "It's a simple issue of home rule," she said. Doyal, along with her husband, sells beer and other goods to the thousands of tourists who pass through their tiny town en route to the Sierra Nevada. "You know that all governmental bodies—including our own county government—operate in a fashion that's best for them, which means they do the most where most of the population is. That has made those of us living in the mountains a stepchild to the politics of the people in the valley."[12]

Bartlett echoed Doyal's assessment of the problem. "They don't understand our lifestyle. They tell us, 'We know better than you.' They give us flatland answers to mountain problems. When they don't ignore us, they push us around."[13]

New-county advocates voiced their interest in local control. Howard "Buster" Ford runs cattle and farms on 2000 acres in the Ponderosa region. He stated, "In a smaller county, chances are most of us would know two or three supervisors by their first names. That's got to be good."[14]

A bit of rural-urban rivalry can be detected in some of the Ponderosa County supporters. Wesley R. Craven, a former cattleman and pack station operator, was a Fresno County supervisor from 1964–72. He told the *Bee*, "We don't like the powers that be sitting in Fresno and telling us we can't do this and have to do that. I think if we got our own county up here we wouldn't rape the mountains like some say that some people are doing. We would be more realistic than the city boys, some of whom have never been here and can't understand the problems that are unique to our area."[15]

Like the natural border at the Tehachapis, many split-county advocates point out, their proposed border at the Friant-Kern Canal is actually a natural boundary line between two distinct geographical, economic, and cultural regions. To the east of the canal rise the Sierra Nevada; the foothill slopes are dotted with brush and live oak, the higher peaks covered with evergreens and granite outcroppings. On the west side of the canal spreads the valley floor, its rich, fertile soils watered by Sierra flows and its landscape dotted by urban centers heated and lit by Sierra hydroelectricity. The Ponderosa region represents about 45 percent of Fresno County's area and 7.3 percent of its assessed value, and contains about 2 percent of its population.

After several months of research and often lively debate among commissioners—two of whom were from the flatlands and two from Ponderosa; one was an outside government analyst—the Ponderosa County Formation Review Commission reached its conclusions. The proposed County of Ponderosa would be able to support the costs of a new government, based on its income from utility taxes, tourism, fishing, mining, and other industries. It would, however, need to rely on Fresno County for its "start-up" costs—which it would repay in time—estimated at $3.3 million. And Ponderosa would have to contract with the existing county for some services, including back-up for its police and sheriff, ambulance and hospital facilities, and some school and prison facilities.

As in any divorce proceeding, the issue that triggered the hottest debate among panel members was how to divvy up the county's assets and debts. Ponderosa County proponents argued that they should not shoulder much of the debt, since services to their side of the county added up to only 1 percent of the whole county's annual budget. They said they receive only $8 million a year in services in return for the $10 million they pay in taxes to Fresno County. Fresno County officials—who eventually came out publicly against the division—argued that because Ponderosa would get the county property now within its borders, it would *owe* the county some money. The two sides hashed it out and finally reached an agreement that each felt was more fair to the other. Using what it called a "fair share" method, the panel based distribution of both assets and liabilities on the Ponderosa area's 7.32 percent of the total county's assessed value. It assessed the new county's worth to be $2.7 million and, because it would get about $3.5 million worth of property, it would owe Fresno County $794,000.

Besides bickering over assets and liabilities, Ponderosa County proponents and opponents argued over what to do with the farmers on the east side of the Friant-Kern Canal. It was decided to include them in the proposed new county. Now living in and receiving the high-quality agricultural services of the nation's number one agricultural county, the farmers of Ponderosa feared they would lose out by being isolated from the majority of Fresno County's farmers. Other residents within the Ponderosa County area also opposed division, calling it "unnecessary duplication" and charging that selfish real estate and development interests were behind the whole movement. Along with support from Fresno County officials and other flatlanders, the "Citizens for Unity" launched a potent counterattack. They focussed on the fear of Ponderosa area residents—many of whom had fled urban life for the peace of the mountains—that a new county might spur unwelcome development in their quiet corner of the Earth.

The Ponderosa County advocates learned an important political lesson from their split-county drive. California law requires that a majority of voters on both sides of the borderline approve the initiative before a county can be divided. But the Mountain Alliance had a next-to-impossible task on their hands: How could they convince flatlanders to vote for the new county? It was clearly not in the flatlanders' interest to do so. Bartlett claimed the main reason Fresno County so vociferously opposed the division was the tax revenues its side would lose if Ponderosa became a separate county. He said Fresno was loathe to lose its "profit center," citing the $10 million Ponderosans chip in to the county coffers each year for only $8 million in services.[16]

Another major obstacle the new county advocates ran into was the flatlanders' concerns about what Ponderosa would do with the water that flows so generously from mountains on the east side of the county west into the valley—for irrigation, industrial, and urban use. Just as Steve Pitcher claimed Southern California would never voluntarily give up "the goose that lays the golden egg," so flatlanders "question whether such an important area [the Sierra watersheds] should be allowed to slip from the jurisdiction of the government unit representing the people who require it," as *Bee* reporter Wanda Coyle put it.[17]

Split-county proponents lost their bid for home rule in June 1982; the Ponderosa initiative lost by a vote of 76 to 24 percent countywide. Because the more populous west side, which has 97.7 percent of Fresno County's population, turned it down by a vote of 77 to 23 percent, the favorable vote Ponderosa area residents cast (52 to 48 percent) got lost in the wash.

The vote proved once again that the less powerful region is hard-pressed to change its underdog status through the democratic process.

Such is the Catch-22 of separationists' attempts to gain local control.

Although they lost, a number of Ponderosa County supporters expressed optimism that the publicity generated by their campaign would bring needed services and improvements to their neglected side of the county. Just as this "PR Effect" aided—at least temporarily—the Yreka rebels' cause back in the 1940s, Doyal said, "the county is treating us better now than they ever have, simply because of the threat of secession."[18] Another interesting lesson future split-staters could learn.

The Effects of a Divided State

Getting back to the topic at hand: many people wonder whether or how California's myriad laws and regulations would be affected by a division of the state. Several would have to be amended to suit the new situation. In addition, federal regulators not now involved in monitoring sales between the northern part of the state and the Southland would suddenly find themselves called upon to cope with a brand new set of interstate transactions. Of course, the feds are amply fortified with existing rules that govern such practices as selling electricity between utilities in different states, setting guidelines for interstate telephone and telegraph rates, and making sure the states don't get too uppitty when it comes to trucking radioactive hospital garments and spent nuclear reactor fuel from one state to be dumped in another. (In 1982, the federal government overturned the State of Washington's short-lived ban on dumping radioactive wastes from out of state, calling it "an illegal restraint on interstate commerce.")

Jerry Bertsch of the California Department of Banking points out that the first question raised by splitting the state—as far as banks go—is the federal ban on interstate branch banking: "The Bank of America, for example, has branches in both Los Angeles and San Francisco." The ban may soon be lifted altogether, but Bertsch doubted in any case that "a split into North California and South California would have any adverse effects on banking. If the law on interstate banking remains, all the companies would have to do is form new corporations. For instance, Bank of America of North California could take all the branches north of the border and B of A of South California could take all the branches south of the line. They have the type of legal talent to deal with such changes."

Another analyst suggested that instead of bifurcating, Wells Fargo, Bank of America, and the other banks with branches in Northern and Southern California would be more likely to lobby either to get multistate banking "grandfathered in" or to accelerate removal of the ban. He predicts it will be lifted within five years.

The Federal Energy Regulatory Commission regulates power sales between utility companies in different states. Jim Harding points out that California already buys power from other states. "If California became two states," he says, "there would be more interstate sales, and FERC would play a more important role in regulating power prices."

More important than increased federal participation, Harding states, would be the specific players involved in regulatory decisions. "It's not what's on the books but who's in power to enforce the laws. For example, if you had an environmentalist in the White House and two bad California governors, you could still do a lot to protect the environment....As for nuclear power's future in either new state, it's hard to say. There's a lot of anti-nuclear sentiment in the North and no decent power plant sites in the South. Diablo Canyon would not be licensed if it came up for initial review today; there are problems along the whole coastline with earthquakes. The feds have already said nuclear power is no business of the states," he said.

A Water Compact?

The biggest problem, without a doubt and without rival, would be water. Even with the state split, the issue wouldn't go away. Will North California try to keep all of its water in the North? What would happen to the State Water Project? And what about all those contracts the Department of Water Resources has with 60 private water agencies and irrigation districts around the state obligating the SWP to deliver 2.2 million acre-feet a year—going up to 4.2 million by 1990?

Some people, the more adamant among separationists, have suggested that the North should keep its water, that most of the 15 million people who live in the arid Southland "shouldn't be there," since the region does not have the resources to support such a population. But Clifford Lee, a water law expert in the state attorney general's office, cautions that it's not so easy to break legally binding contracts, as the SWP contracts to deliver northern water to the South are. "You have to take great care when you discuss abrogation of contracts. That is not to say that certain provisions of contracts might not result in a reduction of water delivered in a certain case." When queried whether or not officials of a new state could simply take over SWP facilities in Northern California, Lee replied hesitantly, "Yes . . . the police power of the state *could* come in and change things."

He points out that the state's contracts are "build-up" contracts, which obligate it to deliver even more water to its contractors than it currently does—most of it to the Metropolitan Water District of Southern California and to the Kern County Water Agency.

"If people don't want a Peripheral Canal or some other facility to deliver more water south, the Department has an out," he stated. Reading from a State Water Project contract, he said, "Subject to the availability of funds, the State will make all reasonable efforts consistent with sound fiscal policies, reasonable construction schedules, and proper operating procedures to complete the project facilities necessary for delivery of project water." In other words, as long as the state keeps trying to complete the SWP—to build a Peripheral Canal or its equivalent—it may not be held liable for failing to deliver the increased amounts of water to which its contractors will soon be entitled, even if the voters don't cooperate.

On the other hand, Lee muses, "A strong legal case could be made that if the water supply contracts were impaired, the bonds [which the state sold to finance construction of the project] would also be impaired. If the project were to deliver less water than it does now, this would reduce the value of these bonds, which are negotiable, like Treasury bonds. This might make the bond-holders unhappy; they could object and sue the state."

A number of people have suggested that the most practical and equitable way to resolve the Two Californias' major bone of contention would be to negotiate an interstate water compact, much like the one that governs use of the Colorado River's flow through seven states and Mexico. In 1977 *New West* insisted that the South will get northern water "because South California has the votes." The writer suggested that an exchange could be made in negotiating the agreement to split the state: "If the South gets a water compact, the North gets the secession votes."[19]

Drawing up an interstate water compact would be no simple chore. The Colorado River Compact is an enormously complex document with several legal loopholes big enough to run a Peripheral Canal through. Because of one of them, Arizona came perilously close to losing its legal claim to its share of the Colorado's flow; California, asserting prior right to the water, attempted to cheat Arizona out of its entitlement. But in 1963, after eleven years in court, the US Supreme Court ruled that nearly 3 million acre-feet of the Colorado River belonged to Arizona.[20]

Interstate agreements, of course, must be approved by Congress—yet another serious obstacle. Clifford Lee dismissed the notion of an interstate water compact as unrealistic and ineffective: "It takes many years to negotiate and get through Congress. What'll that do to current delivery schedules? North California doesn't hold the cards to negotiate [less water deliveries to the South]. South California says it will need the water and it has the contracts for it. It will sue and probably win."

Lee cited the July 1982 Supreme Court decision on the case of *Sporhase* v. *Nebraska*, which stated that the state could not prevent the interstate sale of water within its boundaries. The high court termed it "an illegal restraint on interstate commerce."

Nevertheless, creative North California lawyers would probably find an obscure environmental regulation or law allowing it to reduce its water diversions to the South, thereby setting off a protracted legal battle. In her analysis of Barry Keene's bill, Kate Yeager pointed out, "Protection for the point of origin (county the water comes from) and for the bays and estuaries to which the water would naturally flow if not diverted southward has recently become a major emphasis of the State Water Resources Control Board."[21]

The California Supreme Court is currently reviewing an important, unprecedented case involving the Los Angeles Department of Water and Power's diversions of Mono Basin water. The Department is legally able to divert all of the water in the streams that flow into Mono Lake. Environmentalists are arguing that because the diversions are destroying the ecosystem of the lake, the public trust is being violated. Such an argument has never before been used to try to reduce water diversions and supersede water rights. The outcome of this case could be an important precedent for future water contract disputes between North and South California.

Other laws—such as those protecting endangered species—might be called upon to prevent construction of water facilities or to halt or reduce water deliveries. Of course, the North might try to pass a law invalidating some of the build-up contracts. But when it comes to California's water and who gets it, only two things are certain: anything is possible, and it will be a bloody fight.

The Complexions of the New States

One wonders what the two new states would be like: who would live where? Would people move from one region to the other? Would the states pass radically different laws? Would they bicker constantly over resources or morals? Where would you want to live?

Based on existing demographic patterns, South California—if one included ten counties in the South, with the border lying along the northern boundaries of San Luis Obispo, Kern, and San Bernardino Counties—would be the second most populous state in the Union, with nearly 15 million people. North California would slip to eighth, with 9.6 million. South California would retain about three-fifths of California's wealth, leaving the new northern state with two-fifths.

The ethnic make-up of the two new states would very likely be as distinct as it is now. The Southland is largely Hispanic; at least 1.5 million American citizens of Mexican heritage reside in Los Angeles, as do an estimated half million more illegal immigrants. Blacks and Asians account for the largest ethnic minority groups in Northern California, primarily living in the San Francisco Bay Area. Out of the vast number of Asians living in San Francisco alone, one in 40 is Vietnamese, most of whom immigrated after the Vietnam War. Thus it's very possible that South California would implement further bilingual and bicultural programs and, within a few years, elect a governor with an Hispanic surname. In North California, no one would be too surprised if, after several years, that state's legislature had an Asian and black majority and the governor was fifth-generation Chinese.

Demographic patterns concerning urban-rural ratios would likely shift if California became two states. Now there are more rural people in the North and more urban people in the South. "If California split in two," muses Clyde MacDonald, consultant to the Assembly Water, Parks, and Wildlife Committee, "you'd change current California population dynamics from 75 percent urban dwellers and 25 percent rural to, in North California, 50 percent urban and 50 percent rural. In South California, almost 98 percent would be urban to about 2 percent rural." We'll consider the political implications of these changes in just a minute.

As for each new state's economic base, clearly each would have sufficient wealth to support its own government. In fact, the two new states would likely be healthier than many others. The most reliable forecasters indicate that the economies of both Northern and Southern California look more promising than that of the nation as a whole, in spite of California's recently lowered bond rating.

Both the Southland and the North will benefit from the growth of high technology industries, such as computer systems, microchips, weapons systems, and aerospace engineering. The South will benefit from about a third of this growth, with a larger share going to the Bay Area and a few other parts of the North.

In addition, the South will continue to prosper with its oil drilling and refining, chemical and engineering firms, financial institutions, tourism and recreation, real estate and development, its agriculture, and of course, Hollywood and the thriving television and motion picture industries. In a split-state, new jobs would soon become available in civil service and the many related industries that would spring up to serve the new capital.

North California would also have its agriculture and its government jobs, although Sacramento would probably shrink in size and importance. The North would keep its weapons research and development and biotechnology (genetic engineering), as well as San Francisco's stock exchange, the North's fisheries, timber and paper processing, mining, oil refining, real estate, and tourism, and the booming wine industry.

Per capita income in the Los Angeles basin was $11,287 in 1979. For the Bay Area, the figure was $12,407. In the rural North, however, per capita income was only $8,763. In the next decade, jobs in the Los Angeles basin will increase by 20 percent. San Diego, a fast-growing community that is handling boom-type growth better than Los Angeles did, will experience 30 percent job growth. Employment in the North will climb, but more slowly. The biggest economic problems in the state will be experienced in the rural far north, an area heavily dependent on the timber industry. This industry is severely depressed because of a slumping housing industry and because of the exploitive and short-sighted silvicultural practices of many of the larger timber companies.

The rural far north is surviving, though, and some of the credit must go to California's newest growth industry: marijuana. Innovations in cultivation techniques and the successful adaptation of *sinsemilla*, seedless female plants, to the rainy North Coast environment have led to high yields of potent pot in the hilly, sparsely populated areas of Humboldt and Mendocino counties. The state's crop has been estimated at anywhere from $500 million to $1.2 billion a year. This ranks above lettuce and oranges, and right up with grapes and cotton, as a cash crop.

A number of people have suggested that if the state were split, North California could help support itself first by selling its water to other states including South California at premium prices, then by legalizing marijuana and taxing it.

Frank Gleason says that, while he doesn't support doing so, if the people of North California legalized the cultivation and sale of marijuana and put a 10 percent sales tax on it, they wouldn't have any financial problems at all. People in Sonoma and Mendocino counties, he points out, believe that a state-wide initiative would pass if voters could be shown how much money is to be gained by legalizing pot, taxing it, and thus saving a great deal of taxpayers' money; expensive helicopter surveillance and other costly enforcement methods could be abandoned.

Ecotopia and John Birch Country?

As we saw in chapter 3, split-state folklore portrays the North as a liberal enclave and the South as an unparalleled bastion of conservatism. But we also saw that, while this may have been true at one time, it is no longer so simple. (An aide to Santa Barbara Senator Omer Rains had an interesting comment on an existing distinction between the North and the South's political climate. Tim Hodson said he believed that there is a "qualitative difference between conservatives in Northern California and those in Southern California. In the South, you have wealthy, urban, radical conservatives, *a la* John Birchers. In the North, the rural people are—if you'll excuse the stereotype—'men of the soil.'")

Let's look at how Californians actually voted in recent statewide elections. The counties of the far north tended to vote as a bloc against bond issues and certain social reforms. This could be because they are economically depressed already and don't want to shoulder more taxes, or because these counties historically have felt neglected by Sacramento, or because they tend to be peopled by fairly conservative, rural people and are isolated from the rest of the state by geography, weather, and economy. Many of the counties in the mountains and in the Sierra Nevada foothills, including Fresno County, voted together and often conservatively, though not as consistently as did the far northern counties. Generalities about the rest of the state are harder to make.

There tends to be a split between rural areas and urban/suburban areas, particularly in votes concerning social issues. For example, the rural counties in both Northern and Southern California tend to vote against such things as tax breaks for the elderly and homosexual teachers in public schools. At the same time, the urban and suburban counties of Santa Barbara, Los Angeles, and occasionally even Orange and San Diego, vote along with the only consistently liberal region in the state—the San Francisco Bay Area.

While it is safe to say that some areas, even whole counties, are more liberal or conservative than they are moderate, there is a mix of each in both the North and the South. Thus, statewide elections in the new states would be hard to call. Los Angeles County, with 48.5 percent of the Southland's registered voters, could offset a more conservative vote in Orange, Riverside, and San Diego counties. And the voice of the liberal Bay Area, with 49.7 percent of the Northland's voters, might drown out the more conservative votes of the rural farming and mountain areas of the North. Or vice versa. Clyde MacDonald points out that if the state were split, "San Joaquin Valley farmers and rural people would carry more weight in the North than they now do."

This last possible shift portends ill for the new northern state's environment. In 1982, Assemblyman Mel Levine of West Los Angeles, a liberal environmentalist, proposed a bill to require annual industrial emissions inspections. San Joaquin Valley and Kern County growers opposed it because of their food processing operations. Levine had to exempt some of the agricultural operations just to get the bill out of the assembly.

As for the haunting question of whether the North would really be Ecotopia, let's consider the possibilities. Since Ernest Callenbach's dogmatic novel *Ecotopia* was first published in 1975,* the evidence that "ecotopianism" is indeed taking over the Northland has been piling up. Callenbach says a bank

---

*Unable to convince any New York publisher to risk it, Callenbach published the book himself, under the label of Banyan Tree Books. Within two years, Bantam picked it up, and it has sold steadily—all over the country and the world. US sales are now over 200,000, and *Ecotopia* has been translated into eight languages for sales overseas.

in Walnut Creek uses the word "ecotopia" to describe Northern California, and in San Diego a flossy investment brochure about the North calls it "Ecotopia." The region supports a large number of environmental groups, and awareness of environmental problems is astonishingly high among many different groups and a majority of northern residents. Even in the new tract-house developments in Fremont and Benicia—outlying areas of the Bay Area not considered liberal strongholds—realtors advertise the vanpools available at a particular tract or its short distance from the bus stop or the BART station. Of course, many northern communities—Santa Cruz, Berkeley, most of Marin County, and parts of the North Coast—established their ecotopian loyalties long ago, as did the cow town of Petaluma, with its limits-to-growth regulations.

To refresh readers' memories, *Ecotopia* is the story of an East Coast reporter who, in 1999, sneaks across the international border between the United States and Ecotopia. He finds a newly independent nation in what used to be two and a half states—Washington, Oregon, and Northern California. The book's premise is that the region's mellow, ecology-minded inhabitants got fed up with the reactionary, super-centralized government in Washington and in 1989 violently seceded from the Union. Ecotopia's borders along the Tehachapis and Sierra Nevada are protected by armed guards, and entry by unwelcome Americans is actively resisted. But our hero gets across.

When he reaches San Francisco, the capital, he finds not smog-choked streets and high rises, but rather an environmentalist's dream come true: smiling people are casually riding bicycles, recycle bins dot the streets, and, very occasionally, a small, fuel-efficient car hums slowly by. Curious, he investigates further and discovers that Ecotopia has a stable-state economy, based on renewable resources and self-sufficiency: no cash export crops, no nuclear weapons or nuclear power plants. Instead of breaking their backs to make ends meet, the people work 20-hour weeks and seem to live quite comfortably. They produce and use biodegradable plastics made from plants; their heating and power is supplied by solar, wind, ocean thermal, and small-scale hydroelectric energy; Ecotopians eat only unprocessed, chemical-free food, and travel by walking, bicycling, and using the efficient, low-cost national train system. The government, led by a shrewd woman president, is decentralized: its capital has been broken down into mini-cities where creeks flow instead of traffic and potholes have been planted with flowers.

The Bioregional Perspective

Callenbach, who in 1981 published a "prequel" to *Ecotopia* called *Ecotopia Emerging*, considers himself a "bioregionalist." He asserts that the attitudes

Northern Californians, Oregonians, and Washingtonians have toward their lush, resource-rich environments give them a common denominator more profound and real than the illogical political boundaries that lumped Northern and Southern California together.

"I think Northern California would be a reasonable-size state—about 9 million people," he said recently. "It would be a comfortable-size country or a comfortable state, if we're still a state. It's small enough to have some sense of direct contact between the government and the citizenry. It's small enough that you can travel all over it without greatly exerting yourself and get to know the place. It is a compact entity geographically because of the particular way it's situated, with the drainage all going out through the San Francisco Bay. It makes a whole, gives you a sense that you belong here. You can say, 'This is my place.' It makes some sense."

Referring to the sticky question of how to deal with the water issue, Callenbach said, "I assumed that during the transition period, Northern California would continue to ship water to Southern California, but at realistic prices."

He insists, however, that places like Southern California, without adequate natural resources to sustain themselves, are going to have to start paying attention to the whole question of carrying capacity. "If you don't pay attention to it, you end up paying," he says. "To maintain a city like Los Angeles in the midst of a waterless quasi-desert is an extremely expensive proposition. Of course, it turns out to be terribly unhealthy, too, because of all the smog. In the long run, it isn't reasonable. So in the long run, I assume either that the same thing that happened to the Tigris and Euphrates Valley, or something else drastic, will happen [to Los Angeles]."

This is what the bioregionalists believe, deep in their hearts. Small, but intensely dedicated, groups of activists around the world are seeking to break off debilitating exports of natural resources from their localities and to untie their oppressive bonds to centralized bureaucracies. Bioregionalists view the planet in terms of areas distinguished by climate, culture, physiography, animal and plant geography, natural history, and other descriptive natural sciences. In localities ranging from the Mattole River Valley in Northern California to Samiland in northern Norway, bioregionalists "network" across the miles, exchanging ideas, moral support, and political advice. The most politically active groups lobby for local regulation of their natural resources and retention of their native ethnic cultures.

Bioregionalists in Northern California have been advocating a separate political identity for the region for several years. In 1977, Peter Berg, founder of Planet Drum Foundation in San Francisco, edited a book entitled *Reinhabiting a Separate Country: A Bioregional Anthology of Northern California.* In it, he and other bioregional writers espouse their view that political boundaries will become increasingly irrelevant and bioregional distinctions more and more important as resources become scarcer. To respond to these "inevitable" realities, they urge "devolving" our present "oppressive" social and political structures, reorganizing places according to bioregional lines, and then "reinhabiting" them with sensitivity to the other life forms that share the region. This, they say, would eliminate destructive and wasteful environmental practices and would allow for culturally diverse peoples to live in harmony with their surroundings and their neighbors.

Berg co-authored a piece for *Reinhabiting* with Raymond Dasmann, environmental scientist at the University of California at Santa Cruz. They wrote: "The bioregion that exists largely in what is now called northern California has become visible as a separate whole, and, for purposes of reinhabiting the place, it should have a political identity of its own. It is predictable that as long as it belongs to a larger state it will be subject to southern California's demands on its watersheds....It should be a separate state. As a separate state, the bioregion could redistrict its counties to create watershed governments appropriate to maintaining local life-places. City-county divisions could be resolved on bioregional grounds. Perhaps the greatest advantage of separate statehood would be the opportunity to declare a space for addressing each other as members of a species sharing the planet together and with all the other species."[22]

A few years ago, bioregionalist Jack Forbes, a University of California, Davis, anthropologist and a Powhattan Delaware Indian, devised a plan for dividing California up. Instead of just two states, he separated the five distinct biogeographical provinces of California into five different states. He left the exact borders to be decided later by the people living within each state; and he chose Native American names for them: Palomar, Ramona, Inyo, Yosemite, and Shasta.

Asked why he did not draw exact boundaries, Forbes replied: "You can't just draw some lines and expect people to go along with them. You have to ask the people who live in the various regions where they feel they belong....You have to give Santa Barbara and Ventura the choice whether or not to go with Los Angeles. And Inyo County must be given the choice of joining Nevada." He does think, however, that existing county lines should be used for starters.

Though they don't sound like political lobbyists, many bioregional groups are making headway on the local-government level around the world. In Brittany, Quebec, across the American West, and in New Jersey, pockets of local inhabitants are pressing for control of their historically significant home-lands. Teaming up with local environmentalists, these groups are exerting increasing influence on the centralized governments they ultimately seek to break from.

While Northern California bioregionalists are not currently working to divide the state, their focus on organizing local peoples might just be the missing link in a potential split-state movement—the essential grassroots support so conspicuously absent from recent efforts to divide California into two states.

Support for the Split

Because Northern California is now "the very small tail to a very large kite" and its residents feel powerless in the face of the monolithic South, it will be Northern Californians who will be most ready to lead the battle for dividing the state. Northerners of every stripe feel threatened by the Southland's ever-growing power and by its unending thirst for northern water.

Politicians

The bioregionalists may be the inspiration for the next round of split-state discussions, and they may be instrumental in organizing local peoples, but support from at least a handful of politicians will be crucial to any split-state drive. Prime candidates for such a leadership role would be legislators from both regions who feel frustrated in their efforts to get legislation favorable to their constituents passed.

According to Mitch Stogner, aide to pro-logger Assemblyman Doug Bosco of the North Coast, some southern legislators try to block every land use bill Assemblyman Bosco introduces. Stogner cites the example of a 1982 bill to limit the authority of the secretary of resources on land adjacent to the North Coast's wild and scenic rivers. Bosco's constituents, largely unemployed loggers, felt that the secretary's mandate was too broad and too vague on the use of such lands. They feared that the land would be locked up permanently and the loggers wouldn't be able to harvest the redwoods on them.

The only way Bosco could get his bill passed, Stogner says, was to keep amending it to reassure the "Southern California environmentalists"—primarily liberal legislators from Los Angeles County—that he did not want to build dams on the protected rivers. Rather, his backers merely wanted the right to harvest timber on land next to them. "As a general rule, people south of the Tehachapis don't have a good appreciation of resources issues," Stogner said, echoing the "city boy" complaints of Ponderosa rebels. "They don't understand them—probably because they don't have any resources down there."

Other southern legislators, such as Riverside Democrat Senator Robert Presley, who led the three-year battle to get a statewide vehicle inspection program passed, may be sick and tired of dealing with Northerners who know and care little about the South's problems. John White, consultant to the Assembly Energy and Natural Resources Committee, said that Northern California legislators were not anxious to pass the vehicle inspection bill (which finally became law in 1982). He guessed the reason was that "in the North, air pollution is not considered a worsening problem." Many Northerners, thus, did not want their constituents to have to suffer costly inspections because of the Southland's smog. Other southern legislators with a decidedly more pro-development orientation, such as Paul Carpenter of Orange County, might like to get rid of the environmentalist legislators from Northern California who helped to kill his 1982 bill to exempt the Bolsa Chica wetlands from state coastal regulations.

On the other side of the Tehachapis, northern legislators with more of an environmental bent than Doug Bosco might be ready to split the sheets with the South, though none have come out and admitted this recently. A few years back, however, 14 co-sponsored Barry Keene's bill to split the state.

Keene, 45, a resident of Elk, in Mendocino County, was elected to the California assembly in 1972 and to the state senate in 1978. He represents California's North Coast, from San Francisco Bay to the Oregon border. During his sojourn in the legislature, states his office's release, Keene has authored successful bills to conserve water, safeguard species, restore fishery resources, protect wetlands, and encourage alternative energy sources. A strong advocate of consumer interests as well, Keene wrote the nation's first law on the right to die with dignity and its first comprehensive law on medical malpractice. He is currently leading a broad-based coalition to prevent the Navy from dumping worn-out nuclear submarines off Cape Mendocino.

While in the assembly, Keene offered his bill to create "Alta California" north of the Tehachapis. In introducing it, he stated, "Division of California makes more sense now than ever. If government has grown too big and expensive, let's split it in two. If the bureaucracy has grown too unresponsive and unmanageable, let's shrink it. If decisions are made best closest to the people, let's let Northern and Southern California have their own governments and capitols."[23]

Keene's aide Greg diGiere recently reflected on the motives behind the Alta California bill: "The arguments for it were that if anyone sat down and rationally tried to think of where state boundaries should be, nobody would ever create a state that is what California today is—the lifestyle differences, the economic differences, just the sheer geographic differences. It just doesn't make any sense at all." He admits that the bill was offered primarily "to publicize the water problems." In 1978, legislators were already arguing about the Peripheral Canal. North Coast legislators felt threatened by the prospect of a major new system for shipping even more massive amounts of northern water south.

"Secondarily, the bill was a serious attempt to see if we could generate some support to actually [divide the state] sometime in the future.... There was an effort to get southern support; it didn't succeed," diGiere sighs.

Keene's bill died in the first committee, amidst peals of laughter and sarcastic exchanges between legislators. One wondered "Who would get the State Supreme Court? Who would want it?" The bill didn't even get all the Northern California votes in the committee. DiGiere recalls that only two of the three northern representatives voted for it; "the third, the representative from Sacramento, felt that his constituents had an economic interest in having the capital of the biggest state in their city."

Citizen Leaders

After Keene's bill failed, Governor Brown, several legislators, and other pro-water interests began the drive to pass the Peripheral Canal package in earnest (see chapter 4). While the legislature argued back and forth about the merits and pitfalls of SB 200, the canal bill, and bizarre political alliances sprang up, a handful of Northern Californian provincialists and environmental sympathizers decided a radical solution was the only way to solve the state's endless water squabbles. By the time Governor Brown signed SB 200 into law in the summer of 1980, Doug Carter and his Two Californias Committee were off and running with a petition to "Split the State!" and stop the dreaded canal.

Carter, 39, a jocular former assemblyman from Stockton who looks a bit like Ron Howard, *Happy Days*' Richie Cunningham, and his long-time friend, Leon Pierce, 30, a wholesale jeweler and avid fisherman from San Francisco, spearheaded the Two Californias petition drive. They put out a press release explaining their intentions to gather enough signatures to qualify a measure for the June '82 ballot. Logically, they decided to focus their efforts on voters north of the Tehachapis, their proposed border. (Their border differed somewhat from Keene's proposed line; the Two Californias Committee would divide the state along current county lines, leaving Ventura, Los Angeles, Orange, San Bernardino, Riverside, Imperial, and San Diego counties intact for the southern state.)

Carter and Pierce's major complaint was, of course, the Peripheral Canal, which they argued would destroy the Sacramento-San Joaquin Delta and San Francisco Bay. "I call this the Vietnamization of California," Carter told the *San Francisco Chronicle*. "In Vietnam we had to destroy one part of a country to save another, and that's what is happening in California now."[24]

Exhibiting unmasked Northern California cheek, the Two Californias Committee's press release outlined their strategy: "It is too difficult to explain to Southern California such terms as 'isolated facility,' 'Peripheral Canal,' 'salinity,' 'Eel River protection,' etc. They get bored when conversation begins over water. But, everybody's attention in Southern California will be promptly focussed on an effort to make them a separate state. Once they understand the threat, and we have their undivided attention, we can explain the foregoing terms to them."

Carter asserted that there were a lot of sound reasons to split the state other than the water issue. "People in San Francisco have to pay tolls when they go over bridges, but people in Southern California don't pay tolls on freeways," he said, though both are funded by statewide taxes.

Steve Pitcher, a San Francisco attorney with a legal publishing company, joined the Two Californias Committee that summer and quickly became its chief organizer and researcher. Pitcher, 38, a serious, straightforward man with short dark hair, light grey glasses, and preppy wardrobe, hardly looks the part of a radical, but when he discusses splitting California in two, his eyes betray a hint of his true passion. Pitcher admits it "was kind of a quixotic thing that caught my imagination" back in the '60s when he was at San Jose State. That was when Senator Dolwig was trying to get an amendment passed to divide the state. Pitcher became fascinated with the notion of dividing the state, and he jumped right on the bandwagon when he heard about Carter's efforts.

"What I'm worried about for the future," Pitcher told a slimly attended press conference at San Francisco in November 1980, "is the lack of political independence and the loss of a sense of community, of northernness."

Alas, the Two Californias Committee failed to gather the requisite number of signatures to get the measure on the ballot; Pitcher says they simply ran out of steam—and money. Carter reiterates this: he says he spent about $3000 of his own money to print up the petitions, but "nobody sent any money in."

Pitcher adds that besides lacking the necessary funds to publicize their campaign, "we also needed a clear field. We were up against the anti-canal coalition. They were getting all the money—for a good reason—they were the ones getting out in the hustings. They were driving all the anti-Southern California sentiment into their organization. Virtually we were eclipsed by a more important and practical issue. And it just sort of petered out."

After only six months, the committee dropped its efforts in face of the better organized, better funded, and politically more savvy anti-Peripheral Canal coalition. The Coalition to Stop the Peripheral Canal won the competition for voters' support because it had the advantage of focussing on an immediate, tangible, and stoppable threat, even if their campaign lacked the pizzazz of the Two Californias drive.

Environmentalists

Although the Coalition to Stop the Peripheral Canal was careful to avoid couching its campaign in interregional terms, plenty of evidence suggests that environmentalists themselves—those in the North especially—resent the South's continuing depredations on northern rivers and wildernesses.

While David Nesmith was at Friends of the Earth, telling the press, "It's not a North-South issue; it's whether we want to continue water subsidies for the alfalfa and cotton fields of Chevron Oil," a different group of environmentalists was taking a more gloves-off approach. David Gaines, head of the Mono Lake Committee and now a resident of Lee Vining next to Mono Lake, blames the blithe way Angelenos waste water—by hosing off driveways and sidewalks and indulging in luxuriant landscaping and swimming pools—for the worsening fate of Mono Lake. Where he grew up in Los Angeles, he wrote in *The New Environmental Handbook*, "No one knew or cared very much that the area was really a sub-desert, nor about what was happening at the other end of our taps in the watersheds of the distant Sierra."[25] And David Brower, founder and chair of Friends of the Earth, once told a staff meeting that the Mono Lake campaign was really "about not coveting your neighbor's water."

A split-state movement that focussed on the South's voracious thirst for northern water would attract the sympathy and possibly the support of a huge number of Northern Californians who feel connected to and proud of their lush, natural environment and protective of their watersheds. Northern

California environmental groups and their members would be powerful allies in a split-state drive—*if* they could be persuaded such a drastic move was the only way to stop the South from continuing to prey on the North's resources. That would be easier if Northern California environmentalists came out of the closet with their anti-Southern California bias.

Aside from the environmental groups that proliferate in the North, the region, particularly the San Francisco Bay Area, welcomes all sorts of religious cults and alternative lifestyle groups. The Bay Area's politically potent gay associations and the dozens of affiliations, guilds, unions, societies, and "friends of" elsewhere-voiceless minorities make the Bay Area a popular spot for people with progressive, off-beat and/or unpopular ideas. Such groups and individuals might support a split-state movement to boot the more straight-laced Orange County Rotarians—a common Southland stereotype— out of the state.

That would leave the task of convincing the other 50 percent of the North's voters outside the Bay Area that they would have much to gain from a division of the state. For instance, they'd have much more control over their lives, resources, and beloved environment if the Southland were a separate state and therefore had to formally request the North's favors. Northern California's astounding 92 percent "NO!" vote on the Peripheral Canal suggests that the region, in spite of its widely diverse residents, has the potential for tremendous solidarity and cohesion.

Other Northern Californians who would likely support a split include the farmers and ranchers of the Delta who rely on that body of water's high-quality supplies. This group continues to oppose southern attempts to divert more Delta water south. Of course, as soon as the legislature passes a bill to remove the Eel River from protected status so a dam can be built and more northern water sent gushing along the California Aqueduct, residents in the rural North Coast will be up in arms against Southern California and its Metropolitan Water District.

Northerners who already feel frustrated by the southern-dominated legislature's insensitivity to their needs—for schools, road repair, local programs, and social services—might also be primed to support a division of the state. Teachers in the overcrowded, understaffed public schools may be ready for such a drastic move, asserts Frank Gleason.

"Since Proposition 13, the state has taken a lot of power away from local entities, like school districts [which are reduced to holding cake sales or installing video games to raise essential funds]. The state is now forty-eighth out of fifty in per capita expenditure on students in public schools; I don't think the teachers will work much longer under these conditions," says Gleason. "I think people will get to a point where they feel that they have so little to say about what goes on in Sacramento—it already seems so far away, it's almost like the federal government. People will get so fed up with the whole system that they might want a radical solution; they might be willing to accept splitting the state."

Pot Growers and Smokers

Support for a split-state drive might also come from the legalize-pot camp. Legalizing marijuana, in fact, was one of the principal motives of a quiet, short-lived, split-state effort back in 1980. Around the same time Doug Carter was organizing the Two Californias Committee, a group of people in San Francisco started the Split-the-State Committee. Advocates of local control and minority group representation, the Split-the-State Committee called for a division of the state into "four or five states"—to be "decided by the voters later," according to its press release. After circulating petitions in San Francisco, the committee dropped its campaign in favor of its number one priority—getting a marijuana initiative passed in California. Committee members admit the split-state drive was undertaken largely because they thought a marijuana initiative would pass in the North if it did not have to depend on votes from what they perceived as the more puritanical Southland.

In 1982, the leaders of the marijuana initiative campaign (and formerly of the Split-the-State Committee) were arrested for selling marijuana to help defray campaign expenses. They might be persuaded to reconsider their original plan: first, divide California into two states, then get the northern state legislature and/or voters to legalize pot. It would certainly be less risky to campaigners to approach their goal that way. And they might be able to sway non-smoking voters in these money-tight times that the millions of dollars now spent on ineffective law enforcement could be put to better use, such as improving the education of both teachers and students.

Bad Blood

Anyone attempting to garner support for dividing California in two would be wise to tap into the cultural one-upmanship played by Northern and Southern Californians. California's long and colorful history of North-South dichotomies has given birth to a long list of regional cliches and an enduring competition between the North and the Southland.

As Greg diGiere points out, a profound psycho-sociological phenomenon feeds into the regional biases of Californians. Residents in the less powerful portion always feel obliged to adopt an attitude of superiority. DiGiere says, "Northern California is like a colony, a powerless appendage of Southern California." He says his friends in Southern California grew up with the idea that Northern Californians "grouse and complain, are ungrateful—the traditional colonialist attitude toward the colonized." He believes Northerners "resort to cultural snobbery for the same reasons colonized people do, that 'Dammit! We don't have any power, but we're better people!' "

Feelings of cultural superiority and resentment were widespread among Northerners during the debate over the Peripheral Canal. Threatened by the loss of more of "their" invaluable natural resources and the potential destruction of northern environments, Northerners took to chiding Southern Californians for their wasteful, consumptive lifestyles and their superficial values.

"Southern California is like a spoiled brat that ought to be weaned before it sucks the state dry," wrote a Two Californias Committee supporter from Stanford in 1980. "The southern legislators are a bunch of selfish, self-interested bastards with no regard for the common good of all the citizens of the state," wrote another from Pacific Grove.

Northerners sneered at the Southland's propensity for unlimited growth in spite of extremely limited natural resources. A split-state supporter wrote to the Two Californias Committee: "This seems to be the only way we in the North can hold onto our rights against the aggressive 'growth at all costs' octopus in the South." Such disdain turned to fear in some: "It is my feeling that Southern California will destroy us to satisfy their own needs," a San Franciscan wrote. Along with a $50 donation, an Oakdale man sent the Two Californias Committee a note that read: "I am very concerned that the water thieves will, if allowed, turn Northern California into the same condition that they have the Owens Valley and with no more concern."

We cannot be certain whether or not Northern Californians continue to feel this way now that they have roundly defeated the Peripheral Canal. Steve Pitcher says that if the canal had passed, a Two Californias movement would have "sprung up as a kind of desperation thing—the last gasp before they nail the coffin." He says there's no interest in it now. "Unless and until the legislature outrageously begins to legislate pro-Southern California legislation and tax bills, to such an extent that there's a huge outcry and it comes down to taxation without representation, that kind of galvanization is not going to happen." Yet Pitcher admits that the defeat of the canal is not the final word on water development in California. "I think that the Southerners, at least those in the legislature, are a little chagrined, a little angry at us now because we told them to take their southern-based canal and shove it. That makes politicians angry."

Pitcher is right. Plans are already being laid to try to push through another bill to fund and construct a Peripheral Canal or a through-Delta channel. How far either gets will largely depend on how strongly Northerners feel about their southern neighbors. Surely the historical, cultural stereotypes don't die overnight. And despite the environmentalists' and others' attempts to explain the water issue in rational, economic terms, the North-South split is not to be "explained" away.

"The simple varnished truth is that I detest Lozangeles and everything it stands for," blurted Herb Caen, upon returning to his beloved San Francisco recently. "I can't explain that statement. It's just something that San Franciscans are taught to say from birth. 'If you aren't a good boy we'll send you to Los Angeles.' " (Caen grew up in Sacramento.)[26]

No doubt there are plenty of Northern California residents who still see Los Angeles as a fast-paced, smog-choked, glitter and tinsel capital, full of plastic, nouveau-riche sycophants. Their ability to adapt to the foul environment of the LA Basin, the stereotype goes, attests as much to their perverse sense of priorities as it does to their mutated genetic structure. And, of course, Southern Californians retaliate by calling San Francisco an anachronistic, provincial town, populated by effete snobs who are hung up on Herb Caen, wine-tasting, and new trendy-chic restaurants. Because they see "the City" as overrun by militant gays and lesbians and every breed of whacko that ever existed, Southlanders count their blessings that they live 400 miles away. Whether or not these feelings run deep enough for their owners to actively support a split-state initiative is yet another question. Though they may believe—in their heart of hearts—that these cliches are true, most Americans consider themselves responsible citizens when election day comes; they might be hard-pressed to vote for a division unless they could be convinced it would be beneficial to all parties concerned.

Southern Support

While Northern Californians react with enthusiasm to the idea of bifurcation, Southerners we talked to weren't jazzed at all. The idea of cutting loose those poor fog-chilled snobs in San Francisco hadn't even occurred to them. Why would it?

Still, there are benefits for the Southland. A skilled PR agency could come up with hundreds, thousands, even; here are a few that occurred to us.

Instead of an hour or more flight or an eight- or nine-hour drive up Highway 5, that deadly monotonous asphalt strip that beelines up the scorching Central Valley, Southerners could hop in their cars and bop over to Irvine, Oceanside, or wherever they chose to put their government seat. They wouldn't have as long a wait to see their representatives either, and their legislators would be much more sympathetic and responsive to their needs.

Instead of just two senators from California, four would be elected and sent to Washington to ensure that the 10 percent of the US population that resides in both Californias was being properly represented in our nation's capital. As Jack Forbes says, "We're cheating ourselves by having just one state and two senators. If California was split, Congress would be distinctly different from how it is now."

Gradually, the South California legislature would evolve into a uniquely regional entity and pass laws favoring the Southland's view of the world. Stricter air pollution standards could be implemented and bilingual educational programs would likely flourish.

Forbes feels it is crucial to offer Southern California something of value in exchange for their secession votes. "You have to give them enough local control. For example, the San Diego area is big enough for its own state, and it has always been very independent of Los Angeles. . . . Everyone benefits from more local control," he says. "The size of [legislative] districts would be smaller; more neighborhoods would be represented. More ethnic groups, now excluded, would be represented. It wouldn't necessarily be better representation, but there would be more voices."

Some Southerners would favor a split because they favor water conservation and believe the state's largest consumer of water, agribusiness, won't institute efficient practices unless it's cut off from some of its northern water supplies.

Certain groups of Southern Californians would actually benefit right away. Real estate agents, developers, and construction workers would be gainfully employed in the building of a new state capital. And restaurants, bars, taxis, and other services would provide many new jobs in and around the new capital. The lawyers who would be needed to resolve the many sticky questions of how to divide the bureaucracy and amend state laws would also reap tangible benefits from a division of California.

Ever the optimist, Frank Gleason believes enthusiastically in the potential for southern support for a split. "What you'd have to do is to get developers, investors, and real estate people interested in the possible profits that would be made by building a new capital," he asserts.

Finally, civic-minded South Californians could work in the new state government; many new civil service jobs would open up soon after the split.

But Could It Really Happen?

The job of convincing the state's voters—both in the North and in the South—would be enormous. Support for splitting the state has been weak in both portions in recent decades. The Field Institute, which conducts the California Poll, reported that in 1959, 75 percent of the Californians asked disapproved of the idea of splitting the state; only 9 percent thought it was a good idea. In 1965, when the state legislature was being reapportioned and Southern California was asserting more control in Sacramento, 73 percent of Californians asked still didn't like the idea of dividing the state in two. In 1969 the margin of disapproval shrank to 62 percent, no doubt because northern legislators were unhappy with the actions of the newly reapportioned legislature. That year, more Northern Californians—18 percent—liked the idea than did Southerners—11 percent.

The last time that Field pollsters asked Californians whether or not they wanted to see California become two separate states, in 1981, 72 percent said no, and 18 percent said yes. Although this figure was not broken down into north and south reactions, it is probably safe to assume that with the Peripheral Canal looming large in Northerners' anxiety closet, most of the 18 percent who approved of the idea lived north of the Tehachapis.

The PR agents responsible for drumming up support for a split would have another major obstacle. Most people simply don't take the notion seriously. They feel that the federal government is too entrenched, too dominant, too powerful for local entities to regain lost rights—in spite of Reagan's lip-service to "states rights." And nearly everyone agrees that the political, legal, and practical questions posed by such a prospect are too formidable to even imagine, much less overcome. Still, states have been formed before out of existing states, and while the technical challenges to be met would indeed be great, they are not insurmountable. As Pitcher says, all you need is a good lawyer or two.

The bottom line, then, is whether or not splitting the state is politically feasible. Spencer Michels, producer of KQED's television documentary "Two Californias," points out, "A split state would seem about as likely as a ban on campaign contributions. For even if northern legislators voted for two states, southerners would not, and there are more of them. And a governor would find such a measure difficult to sign, since his support comes from the south and the north. Furthermore, he wouldn't want to preside over a division."[27]

Jerry Brown may have said this best in 1977. When asked his opinion on splitting the state, he said, "Who wants to be governor of a little-bitty state?"[28]

A split-state drive would have to be timed just right. The ill-fated 1980 Two Californias Committee started off with a groundswell of anti-Southland sentiment in the North after Governor Brown signed the Peripheral Canal bill. But the movement lost crucial momentum quickly when environmentalists, Delta interests, and the huge San Joaquin Valley agribusiness firms joined hands to do battle with a tangible enemy—the canal. The committee never gained enough grassroots support; indeed, it never really got organized at all.

No doubt, this was due largely to its severe lack of funds. Frank Gleason estimates that an initiative drive would need about a half million dollars. Add at least a half dozen *experienced* political organizers and a vast network of dedicated volunteers, and a campaign might have a fighting chance against the vigorous counterattack the South would launch. Imagine the Metropolitan Water District flexing its extraordinary public relations muscle on behalf of its own interests and those of southern politicians, southern growers, money institutions, developers, and the myriad industries that depend on Northern California resources. They would start by calling it a wild-eyed, dangerous scheme, and then they would get rough.

Now consider how members of Congress would react. Why would they favor getting two more senators from the Pacific Coast? They probably wouldn't. Clyde MacDonald suggested that "One way to get around that opposition might be to join California with Nevada and then split that into 'Northern Calvada' and 'Southern Calvada.' This would be more appealing to Congress because it wouldn't add any more senators." An intriguing suggestion, to be sure, and somewhat reminiscent of slick Senator William Gwin's 1848 plan to combine California with Nevada and Utah to make one state and then later divide it in two so one could have slavery. Convincing Congress that Two Californias is better than one will be a tall order.

On the other hand, there are people who believe that a division of California is inevitable. Bioregionalist philosopher and activist Peter Berg says, "Northern California is a resources colony. . . . Were the people in Northern California to eventually become both so pinched by depredations being made on this region, and so enthusiastic about assuming a bioregional identity—and I think both things will happen—they would provide the necessary momentum to create a politically autonomous Northern California bioregional government. I think it's inevitable that regions in the biosphere, specifically bioregions, have to become conscious in the minds of the people who live in them, even if they don't want them to be. They don't have to say, 'I want to be a bioregionalist.' They have to say, 'How much am I willing to participate in the destruction of Black Mesa [in the Four Corners area, where a huge coal mine operates on Navajo and Hopi land]? How much am I willing to participate in the diversion of the Columbia River, to get it to Phoenix?'

"Bioregional identity is inevitable from both the positive and the negative side. People are going to be forced into it [because of diminishing resources], and people are going to want to do it."

Ernest Callenbach says, "I wouldn't be a bit surprised" if the state were eventually split in two. "Politics are so volatile these days," he says and cites the example of the Shah of Iran being in power one day and the Ayatollah the next. "Even here it's quite possible; great shocks could happen."

In *Ecotopia Emerging,* Callenbach imagines one set of crises that might touch off a division of the state. "There's a corruption scandal along with a water conflict," he said recently. "Something like that might make people think that the state government is too big, too powerful, too dangerous when it gets corrupted and therefore it ought to be made small. And God knows, the state government was corrupt for so long it's not impossible to think that it could fall under the domination of some interest or other. It's already considerably under the domination of agribusiness. That might become so scandalous that people would be outraged by it, be willing to fight for it. . . .Some issue will come along. . ." his voice trails off.

It's 1986. California politics have become volatile. The governor is self-serving and insensitive to the demands of his constituents, save the big corporations that put him in office. A new political party, "the Survivalists," has been formed to advocate radically different solutions to the state's problems. The Survivalists uncover a secret agreement the governor made with "certain Southern California interests at the time of his election campaign."[39] The agreement involves a large tract of land, as yet undeveloped, along the Southern California coastline and a dam on a Northern California river to supply the water necessary to build on the arid land in the South. (By this time, the Peripheral Canal has been built, and an assault on a northern river is nervously anticipated.)

The scheme promises to net the governor and his partners scads of money. When the newspapers get wind of the plot, the citizens of California pour into the streets.

"In small towns and large cities, masses of people organized by the Survivalists marched in protests bearing signs that read 'Save Our Water! Split the State!' At an all-day rally called by the Survivalist Party in San Francisco's Golden Gate Park, an amazing 150,000 people turned up. And along the hundreds of miles of canal which snaked its way from the Delta southward, and around the giant pumps that consumed more energy than a major city in lifting water over the Tehachapis, state police guards were posted against the possibility of sabotage."[40]

This is Callenbach's vision. It is a Northern California view of things. We might add to the fantasy and look at its other aspects.

Say that it's the second year of a nasty drought. Southlanders have been forced to cut down their use of water: bathing once a week, letting their lawns die, and driving dirty cars. At first they made the sacrifices cheerfully, but as time wore on, the novelty wore off. Now they're sick and tired of living like animals. Their anger flares when they discover that their own water agency, the Metropolitan Water District, has betrayed them. During the campaign for the "new, improved Peripheral Canal," MWD made a behind-the-scenes agreement with its old pal, the Kern County Water Agency. To secure Kern farmers' support for the canal, MWD agreed to give the farmers first rights to any available State Water Project supplies in case of a drought. Now MWD's unwary customers—15 million people in Southern California—are paying. Needless to say, they're angry.

While MWD has been hypocritically exhorting its customers to conserve water, cheap state project water and capital from Central Valley growers and oil companies have rearranged the entire landscape of rural southeastern California. Persistent rumors have it that certain MWD directors are making fortunes off investments in the developments that have dotted the Mojave with golf courses and "Levittowns-in-the-desert." Even the sleepy town of Needles is now a metropolis, lined with palm trees, and sucking water from Northern California. When called upon to explain such use of critical water supplies during a severe drought, an MWD spokesman loses his temper: "We're Metro and we don't have to care," he shouts during a press conference.

This is the final straw. Southlanders hit the streets, crying "Dump the Third California!," "Get MWD and agribusiness off our backs!," and "Grab the water and run!"

Meanwhile, the northern secession has moved along. Radicals bomb the State Water Project's intake pumps at Tracy, disrupting deliveries to the South and the Central Valley. The governor, vowing to protect public and private

NORTHERN   CALIFORNIA

MORE THAN JUST A *STATE* OF MIND

property and "prevent hooliganism," calls out the National Guard units in Stockton and Sacramento.

But these boys are of hardy, Northern California stock, born and bred on Herb Caen and Napa wine. They break ranks and join Survivalist Party groups who are occupying the Delta water facilities. Chants of "Split the State!" and "Save Our Rivers!" are heard echoing through the waterworks.

Shocked Southern leaders realize it's not the time to explain water subsidies and political subterfuge to the people of the Southland. Their water supplies are being threatened by lunatics in the North. They drop their banner to dump the Third California and throw their support behind the governor, vowing to keep the state whole. Angeleno units of the National Guard, called in to replace the traitorous northern units, march north towards the Delta. Joined by angry mobs from cities and rural areas in both parts of the state, the two Guard units meet at the Tehachapis. In a bloody confrontation, half a dozen people are killed and hundreds taken prisoner.

Now the US government steps in. The President sends Marines and regular Army troops to hold the warring factions apart. A temporary no-man's land is established, and an uneasy peace settles in. To prevent further bloodshed, the President calls a statewide election on dividing the state. It passes easily.

Then there takes place what *Time* magazine will later call "the Pakistaniza-tion of California": thousands of Southern families move north to Ecotopia, and certain business leaders, whose careers have been tied up in water exports, drive south, hoping against hope for peaceful old age.

Sounds far-fetched.

Or does it?

How To Split a State

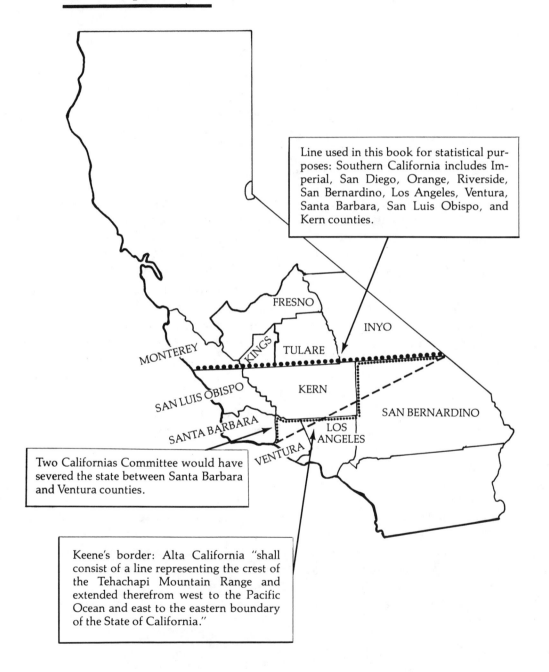

Line used in this book for statistical purposes: Southern California includes Imperial, San Diego, Orange, Riverside, San Bernardino, Los Angeles, Ventura, Santa Barbara, San Luis Obispo, and Kern counties.

Two Californias Committee would have severed the state between Santa Barbara and Ventura counties.

Keene's border: Alta California "shall consist of a line representing the crest of the Tehachapi Mountain Range and extended therefrom west to the Pacific Ocean and east to the eastern boundary of the State of California."

6

California Fusion—
Healing the State

Let us suppose that the decision to split the state of California has been made by the people and ratified by the various necessary legal means. A bilateral commission is established to draw the exact boundary. Where will the boundary fall? For most statistical purposes we have used the ten-county boundary in this book; but there are problems with this line. As the Supreme Court pointed out in 1964, county lines are arbitrary borders of convenience drawn by the state. The decision is a momentous one. Perhaps a better boundary is available.

Let us suppose further that you, the reader, and we, the authors, are selected by the commission to find the perfect border. We start by traveling south from the Oregon border, our senses sharpened, looking for the break.

Out on the Border

As we head south into California from the Oregon border on Highway 101, we pass through redwood country, cross the Klamath River, and reach the town of Eureka. Yreka, capital of the short-lived State of Jefferson, is 100 miles to the east along Interstate 5. Near the Mendocino coast, the highway heads inland through ranching and wine country. Finally we reach the Bay Area, a sprawling, bustling center of five and a half million people. All the concrete is a bit of a shock, but somehow most of the area seems to fit in with what has

> **"If the metropolitan areas in the North and South continue to look at each other as the enemy, we'll never be able to form a rational water policy in this state."**
>
> **—David Nesmith**

199

gone before. Perhaps it is the maritime drift that the Bay brings, San Francisco's spanking whiteness, the big chunks of wild land that have been preserved near the city, like the vast Golden Gate National Recreation Area.

We continue south to the Salinas Valley, an agricultural center. Every town along here seems to be the capital of something: Castroville, the artichoke capital of the world; Brownsville, the begonia capital; Gilroy, the garlic capital. Soon we pass Carmel-Monterey, charming, touristy coastal communities. Life here is good. Perhaps Governor Echeandia would be willing to leave his beloved Josefa and return here if only he could. (Were he still in San Diego, he would probably be proud of the way that city has grown into the twentieth century.)

We leave the Monterey Peninsula behind, and quickly we are due east of Big Sur, Yosemite's rival for the title of California's single most beautiful spot. There lies Esalen, capital of the Human Potential Movement, a California phenomenon whose ranks are swelled equally by people from the North and South.

We stay on Highway 101. Now Hearst Castle is off to the west. We remark to one another that the land is definitely beginning to change. We have entered the transition zone. Writer James Houston describes his impressions of this area.

"I feel it as soon as I squeeze through the rocky gap at Gaviota Pass and swing out onto the coast road below the Santa Ynez Mountains. There is a shift in the landscape, another quality of light. The slopes behind me resemble slopes in Mexico. Where stone shows through the brushy cover, it has the color of bright sand. The air is softer now, and the ocean tropical. Passing through this arm of the Transverse Range, I have crossed some elusive border, and already I hear it coming toward me, the highest octave of a distant early warning system; I feel the outer edges of its irresistible magnetic field.

"I have entered the Southern California continuum . . . ."

Houston tries to help us pin down the borderline. Perhaps, he suggests, it is marked linguistically, where people stop pronouncing the word "rodeo" as rō´-dē-ō and begin to say rō-dā´-ō instead. Or perhaps it is marked by traffic patterns, where the traffic "leaps to the next increment of thickness and something affects the look of the vehicles themselves, the ways they are decorated, the profiles and steering-wheel attitudes of the drivers."[1]

Traditionally, the boundary between North and South has been marked by the Tehachapi Mountains, a range that travelers must cross heading south from Bakersfield, bound on Highway 5 either to LA or toward the Mojave Desert. "The Tehachapi Range," wrote Carey McWilliams, "has long symbolized the division of California into two major regions: North and South." Most observers have agreed. There is debate, though, about the exact point of demarcation along the coast. Some might suggest Hearst Castle itself, legacy of a northern publisher run amuck, Southern California style. Others claim the line is at Point Conception, where the coastline sharply changes direction north of Santa Barbara. Biologist Dennis Breedlove notes that ocean water temperature changes dramatically at that point. Also, many common northern plants, like yerba buena, disappear south of Conception. Ecologist Raymond Dasmann places the line a bit farther north, along the Santa Maria and Cuyama rivers, near the town of Guadalupe.

If you follow the Tehachapis east, they join the Sierra Nevada and run north to Lake Tahoe. As they appear on a map, the mountains seem to make an excellent border. They generally follow county lines, watersheds, planning districts, and even Native American regions.

There are several problems, however, with using the Tehachapis as a borderline, the biggest of which is that the entire Central Valley would belong to the North. The bulk of the draw on northern water would remain North. Of course, voting patterns would be different, but agribusiness would be quick to pressure northern legislators to build giant water projects. The North, strapped for funds without the Southern tax base, might be forced to accede to the demands of such a dominant industry.

A different sort of split is suggested by *Washington Post* writer Joel Garreau in his book *The Nine Nations of North America.* Garreau thinks the North-South split is really between "Ecotopia" and "Mexamerica." Southern California, by this scheme, is part of a large biogeographical province including all of California east of the coast range and south of Sacramento, west Texas, northwestern Mexico, southern Arizona, and most of New Mexico. In Mexamerica, says Garreau, industrial ingenuity is "both father and child to a sense of the miraculous. . . . In Mexamerica, the idea of a freshwater supply flowing unchecked into the sea is considered a crime against nature—a sin."[2]

Northern California, according to Garreau, is part of Ecotopia, a skinny peninsular bioregion that runs from Point Conception to Anchorage, Alaska, never ranging more than 150 miles from the Pacific. What defines Ecotopia? First, it is the only place in the West with enough water. Secondly, people share a mindset that is different from the rest of America. Ecotopia is "sufficiently blessed with resources to inspire thoughts of husbanding what exists, in order to make it last forever. The implication is that others should consider doing the same. . . . In Ecotopia, leaving a river wild and free is viewed as a blow struck for God's original plan for the land."[3]

> We began dividing the Californias just north of San Simeon and cut east-
> ward in a straight line across the top of San Luis Obispo, Tulare, and Kern
> counties through the scrubby foothills of the Sequoias along the edge of the
> village of River Kern. When we reached just south of Chimney Peak, one
> man in our group moved impulsively, turning the boundary at a 90-degree
> angle and starting up the craggy line of Inyo County until the county line
> met Nevada, thereby keeping the Owens Valley for the South. 'This is no
> time to give back what we've stolen,' he said, reaching for a glass of water.[4]
> —New West

Unfortunately, this plan ignores the realities of allegiance out west. Most of the people of northern Ecotopia have little desire to sign on with North California and inherit the travails of a huge urban area like San Francisco. Northern California culture is really hybrid, a blend of California glitter, New England style, and Gold Rush ruggedness. Most Northern Californians consider their hybrid a happy blend and wish to retain it intact while keeping their agricultural base to the east. If San Francisco has a sister city in this country, it might be Santa Fe, a place where opera is big business and where zoning laws require even the Burger King to have terracotta tile floors and open-beam ceilings. Yet Garreau's plan would place Santa Fe in Mexamerica with Los Angeles and Phoenix.

As far as the Southland is concerned, Garreau presumes that Hispanic roots would hold Mexamerica together. But do the people of Southern California, Hispanic or otherwise, feel a closer connection with El Paso, Texas, than with San Francisco, Sacramento, and San Jose? Probably not.

City and Country

This search for the perfect border is a troublesome one. Some would even suggest that the only boundary needed is to separate East California from West California. "California is not divided north and south, as the common myth would hold," writes Tom DeVries in *California* magazine. "It is divided east and west." According to DeVries, West California is a strip some 40 miles wide that runs from Sonoma County, north of San Francisco, to the Mexican border. West California is the exploiter, the taker of resources. East California is the rest of the state, a land of fresh water and open space, "where there are unpaved mountain roads and valleys populated by hicks who irrigate cotton and have grape strikes. Where people are named Clyde, Jose, or Darlene."[5] DeVries argues that urban-rural conflicts are really behind California's problems.

There is something to this argument. Certainly many of the split-state move-ments in California have focused on urban-rural issues. The Yreka Rebellion is a notable example. And people who speak of a firmly united North Cali-fornia are probably out of touch with the antipathy for much of the Bay Area lifestyle that exists in the northern back country. But the idea that California is really divided east-west is a misleading one.

As was stated earlier in the book, nearly every state in America is in some sense divided between city and country, upstate and downstate, capital and cow county. California is burdened with this division; but the more important rift is of a different ilk. It is over the very relationship between civilization and the natural world.

Let us look for a moment at the little community of Round Valley in eastern Mendocino County. This valley contains about 3,000 people. A decade ago it was nearly condemned to a watery grave by a proposed dam on the nearby

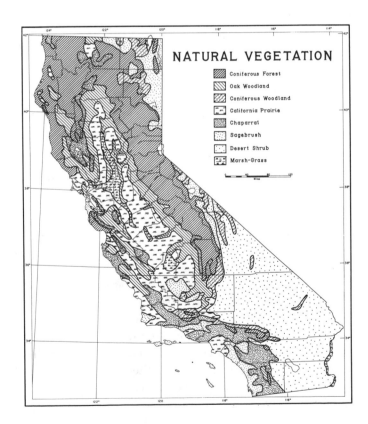

Maps by Arthur Karinen, from California, *Land of Contrast*, © David W. Lantis, published by Kendall/Hunt.

NATIVE GROUPS, 1770

Eel River. Local residents, including members of the large Indian community there, teamed with environmentalists and an unlikely ally named Ronald Reagan to stop the dam. Today, one resident has proposed building a large wood-burning power plant in the valley. A number of local people support this project because it will provide jobs in an area of high unemployment; but another large segment of the community opposes the plant strongly. These people want to maintain the crystal clear air of their special valley. They are not meddling city folk or goldanged environmentalists; they are local people.

There are people named Clyde, Jose, and Darlene living in East California who want to live in harmony with nature, to preserve what David Abelson of the California Planning and Conservation League calls "the rich, integrated, ancient systems" of the natural world.[6] And these rural folk have allies in the metropolitan areas of the North, people named Bruce, Yoko, and Rainbow, people who wish to limit the degree to which the city exploits the natural re-

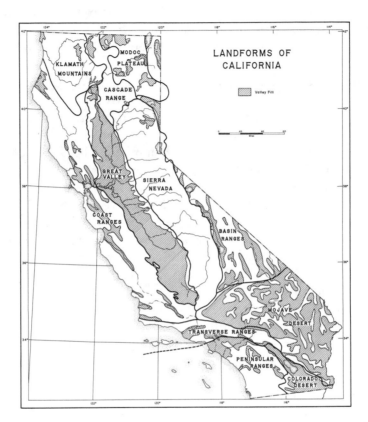

sources of the country. Perhaps this is the key to California's North-South split: a large group of northern city-dwellers are trying to keep the state's cities from devouring the land out back. There may not be another large city in America where this world view is as prevalent.

Lifeblood and Vertebrae

Simply dismissing the idea of an east-west split, though, does not get us much closer to finding the real border. It is important to realize that if a split-state movement does succeed, the decision on border placement would probably be made by a compromise among power-brokers. In that case, we should probably be looking for an equitable border, one that would satisfy our hypothetical bilateral commission. A guess for such a border might be a line perpendicular to California's longer eastern side (which runs at about a 45 degree angle to lines of longitude). Such a line might run from, say, Cambria, just south of Hearst Castle, to Fresno and then on to the Nevada line. Each new state would retain a large chunk of the San Joaquin, although much of the water-intensive cotton crop would go to the South. Bakersfield would be in the South, where it probably would choose to be. Fresno would be right on the border, reflecting the split allegiances of that community. The South would retain the bulk of the Owens Valley; but the North would take charge of Mono Lake and its feeder streams, a locus of great concern to environmentalists. The two states would be close to the same size.

Somehow, though, even this equitable solution to the border problem is not truly satisfying. The placement of any border must inevitably slice through an important north-south running artery and cause the state to bleed. The Two Californias are tied together by all sorts of lifeblood vessels. By the blue Pacific, an ultimate symbol of continuity, a defier of boundaries. By arterial highways, like #1, the Coast Road, a slow-paced magical ride through scenic country; by good, old 101, a pleasant mix of speed and scenery; and by I-5, the speedway, where Porsches and truckloads of tomatoes barrel side by side through the San Joaquin.

The Two Californias are connected by mountain ranges, which—except for the quirky Tehachapis—run north-south like enormous vertebrae, the backbone of the state; by the pipelines and contractual obligations of the state's water system, a colossus of plumbing and engineering; by the earthquake faults, like the famous San Andreas, which course through the state like heartlines, there for the fortune tellers and geologists to read.

To step back and get some perspective on this problem, look at California on a relief map or a globe. It is readily apparent that neither the Tehachapis nor any other feature makes a natural border between Northern and Southern California. In fact, the boundaries of most states make only slightly more sense, are actually embarrassingly arbitrary. The rationale for creating smaller-than-national-level political units is to bring about an entity of manageable size to deal with local problems. This is why the issue of local control has become intermingled with split-state politics.

But really, the issues that threaten to divide California into two states are not local-level issues. They do not involve school district management, police and fire department budgets, local ordinances, and other personal decisions. What divides California is a question of the highest, most general order. The resource in question is water, a basic, flowing, continuous, boundary-crossing resource. The politics of the question involve nothing less than the very power structure of our society. And the philosophical aspects, those of the relationship between civilization and nature, are of the highest order. The matter at hand defies containment in a small decision-making unit.

Imagine, for example, a segment of the San Joaquin Valley near our hypothetical new border. All the farmers in the basin pump water out of the ground for irrigation. Those on the north side of the line are tightly restricted, while the farmers to the south are free to pump as much groundwater as they choose. After a while, the entire groundwater basin, which is contiguous underground and cannot sense our boundary line, will be severely depleted. The problem here is that an artificial border gets in the way of management on an appropriate scale.

It could be argued that even the entire state is too small an area for deciding California's water future. The matter may indeed belong at the federal level. After all, several states are involved directly or indirectly, including all the states in the Colorado River watershed. There have been rumors circulating for years that the MWD and agribusiness may strive ultimately to import water from Alaska or the Columbia River when Northern California resources are stretched too thin or are too politically troublesome.

Much of the time the federal government has served as an ally for the North in the water wars. Several federal agencies lobbied against the Peripheral Canal; federal protection of wild rivers is stronger than that offered by the state; and the infamous 160-acre limitation (now revised) has served in part to restrict the consumption of small farms by agribusiness giants.

Nevertheless, the federal government cannot solve California's water problems. Only Californians can. But how are we to proceed? It seems we have come to the end of the split-state road, taken this proposal as far as we can. Efforts to divide California have served an important function many times over the years. Right now, though, such efforts may be counterproductive. We may need a sharp change of course. With the fabric of our society threatened by powerful forces, perhaps we need to pull the state together more than ever before, to unite movie stars and cowboys, gays and straights, whites, blacks, Hispanics, Asians, American Indians, North, South, East, and West California into one workable version of the Golden State.

Mending the Split

The process of splitting a state is best described as a *fission*, a word defined in the dictionary as: "reproduction by spontaneous division of the body into two or more parts each of which grows into a complete organism;" or, "the splitting of an atomic nucleus resulting in the release of large amounts of en-

ergy." Each of these definitions encompasses part of what split-state advocates
hope will occur if North and South California are created. Secessionists expect
that the energy wasted by needless conflict between the Two Californias could
be used more constructively in a two-state scheme.

The opposite of fission is *fusion.* This word is defined alternatively as "the
merging of diverse elements into a unified whole;" "a political partnership, a
coalition;" and, "the union of atomic nuclei to form heavier nuclei resulting in
the release of enormous quantities of energy when certain light elements
unite."[7] Combining these meanings, we could define fusion as the merging
of diverse elements into a unified political partnership resulting in the release
of enormous quantities of energy. That sounds more like what Californians
may need! Instead of figuring out how to split the state, we should look more
carefully at how to fuse North and South, east and west, city and country,
nature lover and entrepreneur into one California.

How is a California fusion to be accomplished? The first step is to *defuse*
some of the tension between the regions. This means exploding some of the
myths and misunderstandings that have arisen during more than a century and
a half of feuding.

THE THREE MOST PERVASIVE MYTHS ARE:

I. There are irreconcilable cultural differences betwen North and South.

*II. The liberal North has been disenfranchised by the overpopulated and
conservative South.*

*III. Southerners are stealing Northern water and spraying it on their side-
walks.*

In reference to the first myth, we have seen that there are cultural differences
but none of these is momentous enough to cause a split. In fact, Californians
probably have more in common with each other than with people in other
parts of America. Perhaps it is time to say *"vive la difference"* and lay this
myth to rest.

The second myth has several aspects. First of all, it is no longer true that the
South is considerably more conservative than the North. California voting
patterns are so quirky and original anyway that they nearly defy such analysis.
This notion of disenfranchisement stems from the overturning of the Federal Plan
legislature. Could a structural reform of the legislature help heal California?

One idea might be to apportion the state senate on a bioregional basis. The
use of bioregional districts for administrative organization has already achieved
currency in some state agencies. The State Arts Council, for example, is or-

ganized on a bioregional basis. A bioregional senate might be ideal for resource management. It could be argued that the original Supreme Court decisions did not rule out such a plan. The Court said that counties were arbitrary units and lacked prior sovereignty. Neither of these criticisms would apply to bioregions, units established by nature.

State Senator Barry Keene, author of the Alta California bill, is himself beginning to work toward healing the state. He has a proposal that may be the best idea of all about our senate, a body made superfluous by reapportionment. He wants to abolish it. His bill, Senate Constitutional Amendment 51, would establish a unicameral legislature starting in 1990. In this one house, 67 members would serve staggered four-year terms. According to Keene, "The plan will increase accountability by making it impossible for legislators to pass the buck to the other house. It will increase the responsibility for representation by eliminating the interhouse rivalries that too often result in stalemate on important issues." Keene also expects that the change to a unicameral legislature would increase the visibility of the lawmaking process and encourage public participation. As a byproduct, Keene figures that the switch would save at least $23 million per year.

This proposal would not, of course, increase northern representation; but a more visible and responsible legislature could be the North's best insurance against a water deal behind closed doors. Barry Keene has another idea that he thinks might alleviate some of California's North-South tension. He would have the board of directors of the Municipal Water District of Southern California elected by the voters of the area. At present, the directors are appointed by various government agencies. Keene thinks that some of our water problems could be solved by enabling the people of Southern California to make their own decisions more directly.

As constructive as they are, Barry Keene's suggestions are not enough to heal the state. The water problem is too big, too intractable, for a political quick-fix. That brings us to the third and most grave of the split-state myths, namely, that the South is stealing Northern water.

One problem with this myth is that the accuser has dirty hands. Los Angeles is the largest city in America surviving on the importation of distant water. The second largest? San Francisco. So the people of Bay Burg can drink, the Hetch Hetchy Canyon near Yosemite, described by John Muir as one of the most beautiful places on earth, is under water. Some have suggested that the first step in reforming California's water system is to drain Hetch Hetchy; the city could be served by reservoirs downstream.

The other problem with the third myth is that water is not a North-South problem. As we saw in chapter 4, the war over water is being fought among three Californias, not two. It is a corporate farmer, not a Silver Lake sidewalk sprayer, who is most covetous of northern streams. Stripping away the outer layers of illusion from the water myth, however, still leaves us with a core of harsh reality: California has a water problem. A solution to this problem is the key to a California fusion, to making the state whole again.

The Winds of War

In all probability, not one Californian thought the state's water problems were solved by the defeat of the Peripheral Canal in 1982. Jerry Brown was unsure of the proper course, but his successor, George Deukmejian, seems to have some definite ideas. Deukmejian appointed rancher Gordon Van Vleck as secretary of resources, and the latter immediately declared that his first priority was to break the stalemate on water development in California. He was looking, he said, for "ways of sending more water south from Northern California."[8]

What means will be proposed to transport water south? Several influential politicians, including State Senator Ruben Ayala, intend to reintroduce the battered notion of the Peripheral Canal. An alternative to the canal is favored by some leaders of agribusiness. This is called the "through-Delta facility," or the Orlob-Zuchalini Waterway Improvement Plan. It would involve widening existing Delta channels, adding pumping stations, and strengthening some 1000 miles of existing levees in the Delta. Developed by Gerald Orlob, a civil engineering professor at the University of California at Davis, this plan is designed to provide an additional 700,000 acre-feet per year to the SWP without the enormous cost of the canal.

Agribusiness supports this plan because it has a chance to be implemented without the environmental restrictions of the canal legislation. In theory, the full freshwater flow of the Sacramento would actually reach the Delta. This could prevent many of the problems caused by reduced flushing of pollutants. Delta farmers, ferocious opponents of the Peripheral Canal, might support the through-Delta facility because it leaves them less vulnerable to upstream appropriation of their water by southern farmers and urban districts.

Many environmentalists fear, however, that the Orlob Plan would be worse than the Peripheral Canal in its impact on San Francisco Bay. Ronald Robie, head of the state Department of Water Resources, agrees with environmentalists that the through-Delta facility "can only add to [environmental] problems already caused by water project operations in the delta."[9] The problems here are the same reverse flows and saltwater intrusion referred to in chapter 4.

Those who would push for greater water development in California have plans that extend beyond improving the Sacramento Delta system.

The second item on the agenda is a dam on a major north coast river, almost sure to be the Eel, probably along the Middle Fork at Dos Rios, near the afore-mentioned Round Valley, in Mendocino County. The MWD has been some-what cryptic about its interest in the Eel. Farmers in the Central Valley, though, are straightforward. They see an Eel River dam as inevitable. "Eventually," says Stuart Pyle of the Kern County Water Agency, "it will have to be done."[10]

The Eel is a Coast Range river. It heads up a little over 100 miles north of San Francisco and reaches the sea another 150 miles farther to the north, by then the coalescence of four shimmering branches. It is not well-known as a rafting river, but it is loved by fishermen and backpackers. The threatened Middle Fork is the jewel of the system, a summer sapphire tumbling out of the wild and remote Yolla Bolly Mountains. Round Valley residents, led by rancher Richard Wilson, fought to save the Eel in the early 1970s. It was then included under the umbrella of the Wild Rivers Act, but this protection will expire in 1984.

Environmentalists argue that the Eel is too silty for damming; its sediment load is among the highest in the world. This fragile ecosystem has already been damaged by logging and roadbuilding near its banks. The prized summer steelhead population of the Middle Fork would probably be wiped out by a Dos Rios dam. And another wild river, one of the few we have left, would be tamed by the engineers.

Water developers have some support in political circles for an Eel River dam. Reuben Ayala has made several attempts to repeal the Eel's inclusion in the Wild Rivers bill. Legislator Ken Maddy of Fresno is another dam supporter. When Jerry Brown was trying to pass his Peripheral Canal bill, Maddy offered his condititional support. "Give me the Eel," he told the governor, "and you'll get the Canal."[11]

Other projects in the works include a plan to increase the holding capacity of Shasta Dam and enlarging the East Branch of the California Aqueduct. The Shasta plan is a concern to many because of seismic activity and geological instability in the area.

These water development plans, then, amount to one or two large new proj-ects and several extensions of current systems. The total capacity of these additions would undoubtedly amount to several million acre-feet. These changes are necessary, say the developers, to meet an anticipated shortfall of 3 million acre-feet per year in California by the year 2000.

Water Futures: the Environmentalist Agenda

These plans will meet with fierce opposition from environmentalists around the state, particularly in the North. One example of this ferocity is the effort made by a group called Friends of the River to stop the flooding of nine miles of whitewater on the Stanislaus River by the New Melones Dam. When engineers tried to raise the level of the dam in 1979, members of the group—which had been fighting the dam in courts, referenda, the legislature, and Congress for nearly a decade—took turns chaining themselves to boulders along the river, threatening to give up their lives to save the river. Eventually, an agreement was worked out to hold down the level of the lake, but heavy winter rains have flooded the rafting stretches anyway.

Environmentalists, though, are prepared with more than eco-guerrilla tactics. They feel that California can prosper without new water development. They point out that California already uses more than half of the water that comes to the state by natural means each year. (See chart.) Asking for more might well be pushing our luck.

Environmentalists also point out the 3-million-acre-foot (maf) shortfall may be an illusion. The shortfall comes from losing Colorado River water to Arizona under a Supreme Court decision, an expected increase in the population of Southern California, and expansion by agriculture.

It may be that fears of a water shortage in Southern California are exaggerated. Ironically, MWD itself provided ammunition for environmentalists who question the projected shortfall. In January 1981, portfolio managers from Chase Manhattan, Bank of America, and 54 other powerful financial institutions met with MWD directors. The money managers were considering the purchase of MWD bonds. They wanted to know exactly what would happen to MWD if the Peripheral Canal was defeated. MWD assured them that there would be no problem, that contingency plans were ready. These plans included enlarged groundwater storage facilities, installation of low-flow plumbing fixtures in their service area, use of reclaimed sewer water, and other measures.

Now, environmentalists want MWD to follow up on these measures. Even with population growth in the Los Angeles basin, they say, MWD's customers

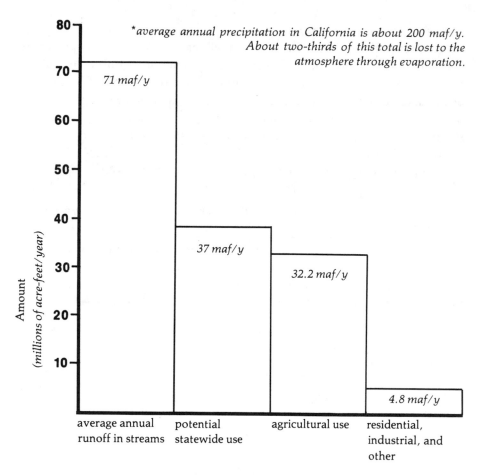

California Water Use

can survive in style with current supplies. In fact, they maintain, Southern Californians are already beginning to conserve water.

Surveys show that the idea of the necessity of water conservation is widely held in both Californias. A recent California Poll shows that 69 percent of Californians feel that it is very important to conserve water. Sixty-eight percent of Southerners held this opinion, a figure that nearly matches the Northern total of 70 percent.

Harry Dennis, author of the book *Water & Power*, believes that Southland water consumption will actually drop in the future. He points out that 45 percent of the water used in single-family dwellings is employed outdoors. The proliferation of condominiums and apartment buildings will reduce consumption figures in the South, as each person will have less outdoor landscape to maintain. Furthermore, all newly constructed homes in California are subject to certain mandated conservation measures, such as required low-flow showerheads and faucets. If any additional conservation measures, voluntary or otherwise, are added to this situation, per capita consumption could drop sharply. Dennis expects the demand of MWD customers to drop to 2.7 maf/y by the year 2000.

METROPOLITAN WATER DISTRICT CUSTOMER DEMAND
(in acre-feet/year)[12]

1980 demand	2000 (according to MWD)	2000 (with mandated conservation)
3,090,000	3,600,000	2,694,100

There may be ways to reduce the need for imported northern water even more. Desalinization of seawater may be one possibility, though it is not cost-effective as yet. Another source is the reclamation of waste water. A Department of Water Resources study shows that about 800,000 acre-feet per year

Thomas Meyer, San Francisco Chronicle © 1982 Special Features

Potential Water Savings in the Year 2000
Through Conservation, Reclamation, and Reuse

SOURCE OF WATER SAVINGS	ACRE-FEET PER YEAR
Interior residential use	
Existing structures	730,000
New construction	510,000
Exterior residential use	200,000
Commercial & Government, interior and exterior	150,000 to 300,000
Urban leak detection and repair	200,000
Industrial use	(unknown)
Agricultural use	1,200,000
TOTAL	Approximately 3,000,000[a]

[a] Since industrial-use savings are not included, this total may be used to represent a lower-bound estimate.

SOURCE: California Department of Water Resources. *Policies and Goals for California Water Management: The Next 20 Years.* DWR/SWRCB Bulletin 4, Public Review Draft, June 1981, p. 46.

could be obtained with current technology. Re-use technology is improving all the time. By the end of this century, reclamation could solve many of our water dilemmas.

All this sounds promising, but remember the magic figure of 85 percent from chapter 4. The lion's share of California water goes for farm use. Any potential shifts in residential use patterns are dwarfed by possible changes in farm water use. A 5 percent reduction in agricultural water use, for example, would amount to a 4.25 percent reduction statewide. This could compensate for one-half of the anticipated 3-maf shortfall.

But how can agricultural water use be reduced? Many experts testify that California farms are already relatively efficient in their habits. There are a few possibilities.

Drastic reduction in per-acre farm water use may come about from the development and use of new technology. Computerized drip irrigation, for example, is 18 times more efficient than sprinkler irrigation and 100 times more efficient than furrows (ditch irrigation). Lasers can be employed to level fields precisely to reduce runoff. Electronic sensors can monitor soil moisture levels at the plant roots, so that water can be applied at just the right time, thus reducing waste. And there is more. Israel has been a pioneer in the field. Surely

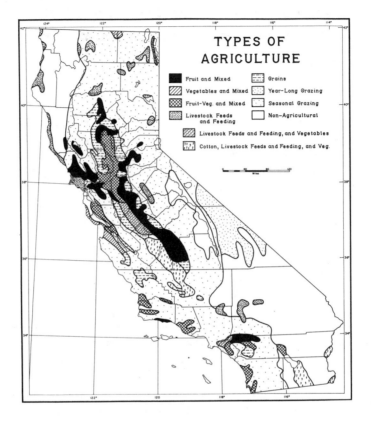

if Israel's agriculture can thrive in a desert, we can make do with the bounty of 40 million acre-feet per year.

The greatest changes in farm water use are available in crop-selection patterns. We grow huge amounts of highly water-intensive crops, such as alfalfa and cotton. Alfalfa requires more than five acre-feet of water per acre of crop. In contrast, a crop like barley needs less than two acre-feet per crop acre. Crops like alfalfa and cotton are very profitable. One of the reasons they are so profitable is that water is so cheap. Water is so cheap because the State of California keeps it that way. There is little incentive for farmers to conserve. If water were more expensive, farmers might decide to install conserving technology, or they might shift their crop-selection patterns. This line of reasoning has led northern environmentalists (with the help of a Southland think-tank) to suggest an interesting solution.

The Price of Change

Several years ago, the California legislature, in a rare display of foresight, commissioned the Rand Corporation of Santa Monica to study water use in the state. The result, a seven-volume opus entitled *Efficient Water Use in California,* appeared in 1978.*

The essence of the Rand Corporation's finding was quite simple: *Let water assume its true cost.* The study found that the cost of farm water in California is artificially suppressed. Four factors help accomplish this: subsidies, property-tax funding, average-cost pricing, and title transfer restriction.

Most Californians have seen those eerie offshore oil drilling rigs in the Santa Barbara Channel and other areas. Revenues from these rigs contribute to the state Tidelands Oil and Gas Fund. Every year, the state legislature allocates about $25 million from this fund for the construction of new water projects. Through 1980, these monies had paid $550 million of the cost of the State Water Project.

The second means of cost-suppression involves the financial structure of water districts like the MWD. These districts supplement their per-gallon tolls with money from property taxes. The average district receives one-quarter of its revenues from property taxes; MWD obtains 33 percent in this manner. This practice sharply skews water prices in favor of farmers.

The highest water use and the lowest water costs in agricultural and outlying residential areas are the result. People in Los Angeles, Santa Monica, Beverly Hills, and other high-tax areas are effectively subsidizing agricultural water use. (The Williamson Act, which freezes property tax on agricultural land, exaggerates this effect even more.) In 1979, for example, the City of Los Angeles received 1.6 percent of MWD's water; yet the people of the city provided 17 percent of the water district's income. Michael Storper and Dr. Richard Walker of the University of California at Berkeley estimate that MWD customers subsidized Kern County agriculture to the tune of $170 million between 1972 and 1979. This sort of fiscal structure distorts the cost and, consequently, the *value* of water.

The third form of disguising water's true cost is average pricing in the State Water Project. When agribusiness contracts for water with the SWP, the charge is based on the average price per acre-foot within the system. As an example

---

*A summary of this report can be obtained from: State Assembly Subcommittee on Water Resources, Committee on Water, Parks, and Wildlife, Room 4130, State Capitol, Sacramento, CA 95814.

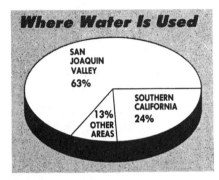

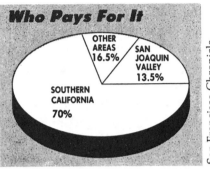

of this, the average cost of state project water in Kern County is about $23 per acre-foot. If the Peripheral Canal had been built, the true cost (based on amortization of expenses and interest) of the water it carried would have been at least $70 per acre-foot (this is according to DWR estimates; other estimates place the projected cost of canal water much higher). Yet the farmers using the water would have paid the average cost, or little more than they had been paying. Every user would have shared the burden of construction costs, which would have been for the benefit of a few large farms, in the most part.

The big farmers are almost compelled to support new construction under this system. Agribusiness would never pay the real cost; society would. In this sense, we are hiding from the tough decisions in our future, hiding in the nostalgic, resource-rich, cheap-water past. As the Rand report bluntly states, "Since the SWP contract charges only the average rather than the marginal cost for capital used in the system, there is an inherent tendency for overexpansion."

The fourth means of cost suppression is restriction of title transfer. In California, the water district, not the user, holds the legal right to water. Some districts prohibit any transfer of a farmer's water allotment; others require that the district broker the transfer. In Kern County, for example, farmers pay for their full allotment; if they conserve water and use less than their allotment, the district sells the surplus at a cheap price and compensates the farmer. But, the farmer ends up paying in part for water conserved. This enables purchase of cheap water by heavy users and penalizes those who save water.

The Rand experts think that all these problems can be overcome by a natural means: the State of California should get out of the water subsidy business. Stop the Tidelands subsidy; fund districts through water-use tolls, not property taxes; use marginal cost, not average cost, pricing; and allow farmers to hold title to the water they are allotted and permit free transfer of water between users.

The theory here is that water prices will reflect the true cost of water to society; the farmers can make intelligent choices on crop selection and water use, including spending money on conservation measures, based on realistically priced water. We are preventing farmers from making socially correct decisions by disguising the cost of water.

WATER PRICES IN CALIFORNIA (per acre-foot)

Agricultural water near Sacramento Delta	25¢–$2
Federal project in Imperial Valley	$4.25
Federal project in San Joaquin	$9.09
SWP in Kern County	$20–30
MWD (Colorado River water)	$50
SWP (supplied to MWD)	$150
Average MWD cost	$100
Average cost to Southern Californians (prop tax included)	$200
Capital cost of all new water projects (average)	$160

Source: California League of Conservation Voters

One area of water use is not addressed by liberating the cost of water, and that area is groundwater. The cost of groundwater is disguised, but not by the state. When a farmer pumps groundwater up in an overused basin, he is adding to the subsequent water costs of each farmer in the basin. The water table drops, and the energy required to pump the next gallon up is increased. In the next few years, the cost of energy will rise dramatically; eventually, pumping costs will become a significant factor in water management. A further problem with groundwater overuse is compaction of the ground itself. Overused basins may become incapable of holding water in the future.*

It is difficult to confront the farmer with the true cost of each groundwater extraction, since that cost is shared subtly by all farmers and will be borne by future generations. The Rand study recommends a pump tax on all extractors. This tax, on a per-acre-foot basis, could help match individual and societal costs. In this case the state would artificially raise the price of water, although the objective would be to allow water prices to reflect "true" cost. This part of

---

*Energy costs may eventually become the overriding factor in determining all water prices in California. Any projects like the Peripheral Canal must be considered in light of the continually rising expense of pumping water over the Tehachapis.

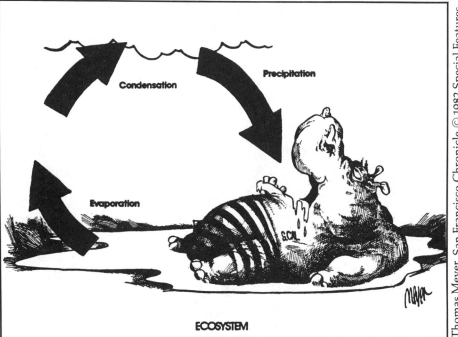

Condensation

Precipitation

Evaporation

ECOSYSTEM

Thomas Meyer, San Francisco Chronicle © 1982 Special Features

the Rand agenda is crucial, because if SWP and other water sources raise prices, farmers might be tempted to seek refuge in artificially cheap groundwater supplies. As was pointed out in chapter 4, however, our already taxed groundwater basins could not survive such a rush.

The environmentalist program for water use in California amounts to this: maximize conservation and efficiency in residential systems; let the price of water reach its true level to maximize conservation (through crop selection, acreage irrigated, and efficient technology) in agriculture; manage our depleted groundwater basins as we would any other precious resource; and hold off on any new water projects until we can gauge the effectiveness of this program. It sounds so logical and simple.

But there is a hitch. A similar proposal, in the form of Proposition 13, the Water Resources Conservation and Efficiency Act, on the November 1982 ballot, was soundly rejected by the voters. Why did this happen? Though it had the surprising endorsement of major newspapers throughout the state, Proposition 13 lacked the campaign funds necessary to educate the public on so complex a series of issues. Its proponents also blame its defeat on misleading tactics by opponents (ads were shown suggesting that passage of the act could result in the construction of the Peripheral Canal!), but problems with this water law run deeper than that. It was not a true *statewide* water plan, no more than was the Peripheral Canal bill, a measure that was also thrashed by the electorate. The kind of legislation that could heal California needs to hear and answer the voices of all the people.

In the next section we will let a few Californians speak for their associates. Representatives will speak for each of the three Californias involved: North, South, and agribusiness. A water wish list of sorts will be drawn for each, in an attempt to cut through the rhetoric of policy-makers and get closer to what the people of the state really want.

The Wishing Well

The North has an overabundance of people qualified to speak for its position on water development—such as Zach Willey of the Environmental Defense Fund, Phillip Leveen of Public Interest Economics West, and David Abelson— but we are looking for a level of meaning that runs deeper than platform or policy. We are trying to discover what it is that the people really want. For that reason, we turn to environmental writer Marc Reisner. His article on Cali-

Thomas Meyer, San Francisco Chronicle © 1982 Special Features

fornia's water problems in *GEO* not only sums up the body of the northern environmentalist program but reveals the heart that beats within.

"There are really two Californias," writes Reisner, "northern California, which has most of the water, and southern California, which has the wealth and power to take it away." Reisner argues reasonably about water development and agricultural use patterns, but like many Northerners, his heart is riding down wild rivers like the Stanislaus. Water to Reisner is not just for drinking and irrigating. Clean, wild water conveys a quality of life that many consider essential. Reisner rejects the position of MWD that freshwater reaching the sea is freshwater wasted. "The price of water development," Reisner maintains, is "the deaths of our rivers. I happen to love free-flowing rivers—I consider them among the most magical things in nature. And the stilling of another singing river by another silent reservoir makes me want to cry. I am prepared to accept reservoirs for the benefit of the people. But the time when reservoirs in California contributed to the welfare of people ended years ago. Now they contribute mainly to the welfare of agribusiness—and the perpetuation of waste."[13]

Reisner's eloquent testimony refutes claims that Northern California opposition to water exportation is selfish, xenophobic, or based on a "don't go near our waterhole" reaction (see chapter 4). The people up North who work the hardest to prevent big water projects are, in many cases, indulging an affair of the heart with nature, with wild rivers, lonely estuaries, shimmering lakes. Theirs is an instinct of preservation, of helping a crucial element of life survive on earth.

In condensed form, a wish list of Northern California water reform priorities would include the following:

1. The ecosystem of the Sacramento Delta and San Francisco Bay must be restored to good health; no further degradation is permissible.

2. The reduction of water in Mono Lake must be stopped.

3. The Eel River must not be dammed; it should receive legal protection as a wild river in perpetuity; other north coast rivers should be similarly protected.

4. The state's groundwater reserves must not be depleted.

The question of a representative for the South involves some of the same considerations as did the selection of Marc Reisner. Qualified "official" speakers abound, such as David Kennedy, assistant general manager of the Metropolitan Water District of Southern California (MWD). Such a speaker could recite percentages and consumptions of millions of acre-feet and draw population growth curves and prove conclusively that we must have a Peripheral Canal or something like it. But we want to get closer to the heart of the matter. What do the people really want?

A most touching and evocative statement of the Southern California position turns up in Joan Didion's *The White Album*. In one essay, she describes a visit to the computer control room of the State Water Project, with its awesome power to electronically manipulate millions of gallons of water daily. She communicates a real excitement about the technology and science of making people feel safe and comfortable.

Didion cautions that water has only recently come under human control and that water is the only natural force we do control in the arid west: "The apparent ease of California life is an illusion, and those who believe the illusion real live here in only the most temporary way. I know as well as the next person that there is considerable transcendent value in a river running wild and undammed, a river running free over granite, but I have also lived beneath such a river when it was running in flood, and gone without showers when it was running dry."[14]

As some label northern conservation efforts selfish and territorial, so Southern Californians have been maligned as profligate water wasters, sidewalk washers, and swimming pool junkies. There are probably water wasters in the South, just as there are undoubtedly selfish or small-minded people in the North. The point is, however, that behind both the Northerner's impulse to conserve and the Southland wish to assure sufficient supplies are legitimate human needs that must be taken into account in water management plans.

Joan Didion reminds us that each of the Southland's "residential water-consumption units" contain living, breathing human beings, with inalienable rights in our society and fully deserving of a high quality of life.

A distilled water wish list for the Southland might look like this:

1. We need reasonably priced, clean, and potable water.

2. We need emergency reserves available in case of a serious drought.

3. We want to keep the conveniences we have won already; we want to take showers, water lawns, even fill swimming pools.

Based on these wish lists, it appears that northern views are motivated very strongly by a concern for the environment, a desire to preserve the very hospitability of the land that attracted many to the North in the first place. In the South, water wishes are motivated equally strongly by anxiety that the inhospitability of Southern California land could become a problem without sufficient water reserves. Any water plan that would achieve statewide acceptance needs to address these motivations.

What about the Third California? Getting to its heart is a different problem. Who are California's farmers? They are, in part, long-time tillers of the soil, dedicated, hard-working Californians. They are also corporations housed in steel and glass, oil companies, communications conglomerates, and other titans of industry. If you listen carefully, though, you can pick up the real motivation of farm water policy in the words of even corporate spokespersons.

Nothing is as important as water to farmers. Technology has enabled agribusiness to develop almost any kind of seed, fertilizer, or pesticide. Giant machines can turn the land upside down or inside out. Crops can be grown even on the most forsaken-looking soil. Almost anything can be provided by the modern farmer—except water. "You can make gasoline out of cow manure if you have to," says one western farmer, "but you can't make water." In the colorful phrase of one western planner, water is "the testicles of the universe."[15]

California has surely gone out of its way to bring water to farmers, thereby making agriculture the state's number one industry. What are the farmers worried about? Listen to William Du Bois, Director of Natural Resources for the California Farm Bureau Federation, as he spoke to a state congressional hearing on California's water future in July 1982.

"It is obvious," Du Bois said, "that it's not possible to serve an ever-increasing population and economy with adequate water while at the same time maintaining and enhancing optimum natural habitat conditions for wildlife. We believe reasonable efforts are warranted to maintain wildlife, but that the basic needs of people are the highest priority for the use of water. . . . There is no status quo for the future of agricultural irrigation. As cities take over farmland, they get the water to match. . . . One thing we must all keep in mind is that *they* are after us."[16]

Who are *they*? Who is the enemy of the California farmer? Attorney John Penn Lee is officially authorized to speak on political issues for agribusiness giants J. G. Boswell Co. and Salyer Land Co. Before the Peripheral Canal election, Lee spoke of his companies' anxieties about water supplies under the present system: "Southern California has the political clout to take water away from plants and give it to people. And if there's another drought, and there's a question of whether it's going to be the plants or the people, you know what's going to happen."[17]

So it seems that California agriculture is most concerned about its own political ally, the people—or rather the growing population—of Southern California. As big water projects have become more difficult to implement, as California loses some of its Colorado River entitlement, as the state's population continues to grow, agribusiness is beginning to feel squeezed. The irony is that agribusiness may not be able to start any new water projects without help from Southland voters; yet it is those very voters that agribusiness is competing with, or thinks it is competing with, for the water.

If this is all a bit confusing, let us clarify it with a water wish list for agribusiness:

1. We need enough water to maintain the present status of agriculture as California's leading industry and of California as the nation's leading farm state.

2. We need to count on this water regardless of population growth in urban areas.

3. We do not want to be told where to grow, what to grow, or how to grow our crops.

There is one other water wish, by the way, and it seems to transcend regional lines. Nearly all Californians agree that *all future California water projects must be fiscally responsible.* This belief was underlined by the dramatic defeat of the Peripheral Canal. Exit polling showed that 65 percent of those Southern Californians who voted against the canal did so *primarily* because of its expected high costs. Even in the North, where arguments against "the dreadful ditch" ran rampant, excessive cost was given as the leading argument by 34 percent of the nay-sayers. Pro-development water forces are still powerful, but their blank check has been voided.

Healing Waters

What do these wish lists reveal? First, they reveal that Northerners, Southerners, and farmers all have something in common: they are all anxious about the future. Secondly, a glimmer of hope appears about putting together a comprehensive water plan. Here are some possible approaches:

• Some version of the Rand plan could be implemented, perhaps on a trial basis, say, for 20 years.

• Subsidy monies could be made available to farmers as low-interest loans for installation of conservation devices.

• If feasible geologically, the further filling of the Shasta Dam reservoir could be permitted.

• A constitutional guarantee of water availability could be made for every

Californian, within the limits of new growth guidelines. How many people is California willing to provide for? Fifty million? Thirty million? This provision could include limitations for population growth within areas served by the State Water Project.

Population growth is the real issue. The North is afraid of the South growing too big to water; the farmers are afraid the South will grow up and take their water; even the Southlanders are afraid of their own growth. We have the water to handle the people who are here now and a modest future growth. If growth is unlimited, if any of the booms of the late nineteenth and early twentieth centuries are repeated, who knows what will happen?

This kind of comprehensive planning will be very difficult to accomplish. A century and a half of feuding will have to be put aside. But the alternatives are bleak. On one hand, we face water shortages, rationing, a crippled farm industry. On the other hand, we face destructive, expensive boondoggles, dying rivers and estuaries, and a state more thoroughly divided than ever before in its history.

There is hope, however. Most of the western states have comprehensive water planning. Arizona, for example, enacted such a law in 1980. This bill required the registration of all groundwater pumping facilities, created a free market for buying and selling water rights, and empowered the state to set per capita consumption limits for urban areas.

Perhaps some day in the not-too-distant future, a northern and a southern member of a newly reorganized unicameral assembly will cosponsor a statewide water plan, a plan that provides guarantees to all citizens of sufficient water, that allows for a propserous farming industry, and, most importantly, that protects a fragile and beautiful California environment. If that day comes, just as the bill is signed into law, the Golden State's oldest feud will come to a peaceful end.

The Wave of the Future

What would the state be like after a California fusion? Would there be "a release of enormous quantities of energy"? One interesting result might be an end to the politics of scapegoating. We could stop blaming each other—Northerners, Southerners, straights, gays, whites, minorities—and get on with the business that lies in our future.

We face some difficult choices. The budget needs balancing, just as our schools, libraries, and poor people cry out for a place on the priority list. We are at a crossroads in coastal development. Is the coast one of our most prized public resources or simply our most valuable real estate? What about oil exploration? How many barrels of crude is a shorebird worth? Does nuclear power have a role in our energy future? How should our society adapt to the increasing numbers of Hispanics, Asians, blacks, and other minorities living in California? A united state would be so much better equipped to deal with these issues than would a house divided.

In fact, a united California could be a very interesting place. A cooperative high-tech industry could set the pace for a recovering economy in America. A new, highly coordinated statewide entertainment industry might develop, using the talent and natural resources of both North and South. Perhaps the Raiders are the wave of the future. All California sports teams could commute by air, playing home games in both regions.

But wait. We're getting carried away here. A united California need not be a homogeneous one, and that is probably just as well. Some elements of cultural exchange are truly unimaginable, like Herb Caen living in the San Fernando Valley and driving the freeways for three or four hours each day. Or like Steve Garvey in Marin, sharing a hot tub with a whole party of laid-back types. Some things are just not meant to be. We can put a man on the moon. We might be able to solve an excruciatingly complex water problem. But nothing on earth can turn a dyed-in-the-wool San Francisco Giants rooter into a fan of Mr. O'Malley's Dodgers. *Vive la difference!*

THE END

Notes

1. The Golden State 1820–1914

1. Lavender, page 41.
2. Guinn, page 223.
3. Beck and Williams, page 121.
4. Cleland, pages 62–63.
5. Lavender, page 30.
6. Beck and Williams, page 123.
7. Kirsch and Murphy, pages 293–294
8. Ibid., page 296.
9. Ibid., pages 301–302.
10. Kirsch and Murphy, page 309.
11. Lavender, pages 31–32.
12. Ellison, page 4.
13. Rush, page 14.
14. Beck and Williams, page 156.
15. Delehanty, page 59.
16. Watkins, page 72.
17. McWilliams (1946), page 62.
18. Ellison, pages 17–25.
19. Ibid., page 26.
20. Guinn, page 229.
21. Ellison, page 40.
22. Ibid., page 56.
23. Latham, pages 125–132.
24. McWilliams (1946), page 116.
25. Kelley, page 157.
26. *Los Angeles Express*, April 12, 1880.
27. *Sacramento Union*, February 8, 1881.
28. Willard, pages 22–23.
29. McWilliams (1946), page 150.
30. McDow (1970), page 41.

2. Los Angeles Rising— 1914 to 1982

1. McDow (1970), page 43.
2. *San Francisco Chronicle*, September 15, 1918.
3. McWilliams (1946), page 129.
4. Ibid., page 135.
5. Ibid., page 136.
6. Lillard, page 223.
7. *San Francisco Chronicle*, November 28, 1941.
8. *U.S. News and World Report*, February 8, 1965, page 62.
9. *Santa Barbara News-Press*, October 28, 1960.
10. McWilliams (1949), page 211.
11. *U.S. News*, February 8, 1965.
12. *Los Angeles Times*, January 4, 1965.
13. McDow (1966), page 33.
14. Brown, Sr., Edmund J., page 212.
15. *Sacramento*, February 1981.

3. A Tale of Two Cultures

1. *Holiday*, October 1965.
2. *Saturday Review*, October 30, 1943, page 3.
3. Garreau, page 5.
4. Houston, page 203, quoted by permission of the publisher.
5. Caen, *San Francisco Chronicle*, November 18, 1982 © San Francisco Chronicle, 1982. Reprinted by permission.

6. Theroux, *California and Other States of Grace*.

7. *Time*, 1966.

7a. McCoy, page 155.

8. *New West*, January 1981, page 54.

9. Ibid., page 58.

10. Holt, page 57.

11. *San Francisco Chronicle*, November 4, 1982.

12. *Holiday*, October 1965.

13. Didion (1970), pages 13–14.

14. California Department of Transportation, "Travel and Related Factors in California," Annual Summary, 1981.

15. *The New Yorker*, September 20, 1982.

16. Dunne, page 10.

17. Houston, page 205.

18. Brown, page 213.

19. *San Francisco Chronicle*, June 6, 1982.

20. Ibid., October 1, 1982.

4. Water Politics and the Third California

1. Jelinek, page 4.

2. Ibid.

3. Villarejo, page 5.

4. Kahrl (1978), page 21.

5. Reisner, page 114.

6. Cleland and Hardy, page 61.

7. Jelinek, page 31.

8. Reisner, page 115.

9. Jelinek, page 64.

10. Kahrl (1982), page 2

11. Ibid., page 3–4

12. McWilliams (1946), page 184.

13. Kahrl (1982), pages 15 and 18.

14. Ibid., page 19.

15. Ibid., page 20.

16. Ibid.

17. Ibid., pages 83–85.

18. Kahrl (1978), page 33.

19. Beck and Williams, page 309.

20. Wiley and Gottlieb, page 109.

21. *San Francisco Chronicle*, June 1, 1982.

22. Ibid., May 20, 1982.

23. Ibid.

24. Ibid.

25. Ibid.

26. Ibid.

27. *New West*, June 6, 1980.

28. Ibid.

29. Ibid.

30. Ibid., page 65.

31. *New West*, September 10, 1979.

32. Ibid., page 42.

33. *San Francisco Chronicle*, May 20, 1982.

34. Ibid.

35. Ibid., June 9, 1982.

36. Ibid., May 11, 1982.

37. *California Living*, April 11, 1982.

38. *San Francisco Chronicle*, June 9, 1982.

39. Ibid.

40. Kahrl (1978), page 58.

41. Ibid.

42. Ibid., page 62.

43. Dennis, page 131.

44. Ibid., page 132.

45. Baker and DeVries, page 39.

46. Ibid., page 40.

47. Dennis, page 121.

48. Ibid., page 122.

49. Ibid.

50. Reisner, page 119.

51. Kahrl (1978), page 108.

52. Dennis, page 43.

53. *San Francisco Chronicle,*
    October 25, 1982.
54. Ibid., April 26, 1982.
55. Kahrl (1982), page 434.
56. Kirsch, "Politics of Water."

5. Imagine Two Californias

1. Latham, page 127.
2. Widney, page 23.
3. Sheridan, page 1.
4. Ibid.
5. Ibid., page 2.
6. Ibid.
7. Latham, page 128.
8. Lawren, page 89.
9. Yeager, page 2.
10. *New West*, May 9, 1977.
11. *Fresno Bee*, May 2, 1982.
12. *San Francisco Chronicle,*
    May 2, 1982.
13. *Fresno Bee*, May 2, 1982.
14. Ibid., May 3, 1982.
15. Ibid.
16. Ibid., May 4, 1982.
17. Ibid., May 3, 1982.
18. *San Francisco Chronicle,*
    May 24, 1982.
19. *New West*, May 9, 1977.
20. Reisner, page 113.
21. Yeager, page 2.
22. Berg and Dasmann, page 7.
23. Keene, press release.
24. *San Francisco Chronicle,*
    October 1, 1980.
25. DeBell, page 90.

26. *San Francisco Chronicle,*
    November 11, 1982.
27. *California Living,* April 11, 1982.
38. *New West*, May 9, 1977.
39. Callenbach (1981), page 286, ©1981
    by Ernest Callenbach, quoted by
    permission of the author.
40. Ibid., page 288.

6. California Fusion— Healing the State

1. Houston, page 198, quoted by
   permission of the publisher.
2. Garreau, page 5.
3. Ibid.
4. *New West*, May 9, 1977.
5. DeVries, page 180.
6. Powledge, page 715.
7. *Webster's New Collegiate
   Dictionary* (1980).
8. *San Francisco Chronicle,*
   January 31, 1983, page 5.
9. *San Francisco Chronicle,*
   June 3, 1982, page 3.
10. Baker and DeVries, page 40.
11. Kirsch, page 70.
12. Dennis, pages 89–91.
13. Reisner, pages 120–122.
14. Didion (1979), pages 64–65, quoted
    by permission
15. Garreau, pages 4 and 311.
16. California's Water Future,
    pages 114 and 117.
17. Kirsch, page 68.

Bibliography

The books we found most valuable in our research and for their intrinsic value are the following. They are referred to throughout the text.

Warren A. Beck and David A. Williams's *California: A History of the Golden State,* for general history.

Joel Garreau's *The Nine Nations of North America,* for an understanding of the nature of interregional conflict.

James Houston's *Californians* and Carey McWilliams' *Southern California Country: An Island on the Land,* both giving a feel for the state's people and cultures.

Lawrence J. Jelinek's *Harvest Empire: A History of California Agriculture,* for a concise and intriguing discussion of the development of modern agriculture in California.

William Kahrl's *California Water Atlas,* the best single and most beautiful and valuable source on California water resources and water development history.

Peter Wiley and Robert Gottlieb's *Empires in the Sun: The Rise of the American West,* for hard-hitting political history hard to find elsewhere.

1. The Golden State, 1820–1914

Bean, Walton. *California: An Interpretive History.* New York: McGraw Hill Book Company, 1968 and 1973.

Beck, Warren A. and David A. Williams. *California: A History of the Golden State.* Garden City, New York: Doubleday & Company, 1972.

Campbell, Horace. *A Short History of California.* Philadelphia: Dorrance and Company, 1949.

Cleland, Robert Glass. *From Wilderness to Empire: A History of California.* New York: Knopf, 1962.

Delehanty, Hugh J. "The California Split." *Sacramento,* February 1981.

Ellison, William Henry. "The Movement for State Division in California 1849–1860." (Masters thesis, University of California, Berkeley, 1913).

Fracchia, Charles. "The Political Faultline." *Air California Magazine,* August 1978.

Guinn, J. M. "How California Escaped State Division." *Annual Publication of the Historical Society of Southern California,* 1905, Los Angeles: George Rice and Sons, 1906.

Hutchinson, W. H. *California: Two Centuries of Man, Land, and Growth.* Palo Alto: American West Publishing Company, 1969.

Jelinek, Lawrence J. *Harvest Empire: A History of California Agriculture.* San Francisco: Boyd and Fraser Publishing Company, 1979.

Kelley, Robert L. *Gold vs. Grain, The Mining Debris Controversy.* Glendale, Calif.: The Arthur H. Clark Company, 1959.

Kirsch, Robert and William S. Murphy. *West of the West.* New York: E. P. Dutton & Company, 1967.

Latham, Milton S. "Message from the Governor." *Journal of the Assembly, California 1860.*

Lavender, David S. *California: A Bicentennial History.* New York: W. W. Norton, 1976.

Los Angeles Evening Express, many dates in 1880 and 1881.

McDow, Roberta M. "State Separation Schemes, 1907–1921." *California Historical Society Quarterly,* Spring 1970.

McDow, Roberta M. "To Divide or Not to Divide?" *The Pacific Historian,* Autumn, 1966.

McWilliams, Carey. *Southern California Country: An Island on the Land.* New York: Duell, Sloan & Pearce, 1946.

Rush, Philip S. *The Californias 1846–1957.* San Diego: P. S. Rush, 1957.

Sacramento Daily Record Union, many dates in 1880 and 1881.

Watkins, T. H. *California: An Illustrated History.* Palo Alto: American West Publishing Company, 1973.

Widney, J. P. "A Historical Sketch of the Movement for a Political Separation of the Two Californias, Northern and Southern, under both the Spanish and American Regimes." *Annual Publication of the Historical Society of Southern California 1888–9,* Vol. I, Los Angeles: Frank Cobler, "The Plain Printer," 1889.

Wiley, Peter and Robert Gottlieb. *Empires in the Sun: The Rise of the American West.* New York: G. P. Putnam's Sons, 1982.

Willard, Charles Dwight. *The Herald's History of Los Angeles City.* Los Angeles: Kingsley-Barnew and Neuner Company, Publishers, 1901.

2. Los Angeles Rising, 1914–1982

Brown, Sr., Edmund G. *Reagan and Reality: The Two Californias.* New York: Praeger Publishers, 1970.

Conmy, Peter Thomas. "Division of the State of California." *The Native Son,* October–November 1980.

Delaplane, Stanton. Articles in the *San Francisco Chonicle.* November 29–December 10, 1941.

Lee, Eugene C., ed. *The California Governmental Process.* Boston: Little, Brown & Co., 1966.

Lillard, Richard G. *Eden in Jeopardy: the Southern California Experience,* New York: Alfred A. Knopf, 1966.

McWilliams, Carey. *California, The Great Exception.* New York: Current Books, 1949.

Marine, Gene. "Should We Bust Up California?" *Frontier,* November 1964.

Reinhardt, Richard. "The Short, Happy History of the State of Jefferson." *The American West*, May 1972.

Salzman, Ed and Ann Leigh Brown. *The Cartoon History of California Politics*. Sacramento: California Journal Press, 1978.

U.S. News and World Report. "Should California Be Chopped in Half?" February 8, 1965.

3. A Tale of Two Cultures

Albright, Thomas. Article in the *San Francisco Chronicle*. November 4, 1982.

Burdick, Eugene. "The Three Californias." *Holiday*, October 1965.

Caen, Herb. *San Francisco Chronicle*.

"California Artists." *New West*, January 1981.

California Opinion Index. San Francisco: The Field Institute.

"California—Sources and Resources." *Saturday Review*, October 30, 1943.

"California Split." *New West*, May 9, 1977.

Didion, Joan. *Play It As It Lays*. New York: Farrar, Straus & Giroux, 1970.

Dunne, John Gregory. "Chinatowns." *The New York Times Book Review*, October 25, 1982.

Holt, Patricia. *San Francisco Chronicle*. January 21, 1983.

Houston, James D. *Californians: Searching for the Golden State*. New York: Alfred A. Knopf, 1982.

McCoy, Esther. *Five California Architects*. New York: Reinhold Publishing, 1960.

Michels, Spencer. "Two Californias." *Focus*, May 1981.

Morgan, Neil. *The California Syndrome*. Englewood Cliffs, N.J.: Prentice-Hall, 1969.

Theroux, Phyllis. *California and Other States of Grace*. New York: William Morrow, 1980.

4. Water Politics and the Third California

Baker, George L. and Tom DeVries. "How the Big Oil Companies are Grabbing California's Water." *New West*, June 16, 1980.

Beck, Warren and David A. Williams. *California: A History of the Golden State*. Garden City, New York: Doubleday & Company, 1972.

"California Opinion Index: A Survey of June Primary Election Voters." San Francisco: The Field Institute, July 1982.

Cleland, Robert Glass and Osgood Hardy. *March of Industry*. Los Angeles: Powell Publishing Company, 1929.

Dasmann, Raymond F. *California's Changing Environment*. San Francisco: Boyd and Fraser, 1981.

Dennis, Harry. *Water & Power: The Peripheral Canal and Its Alternatives*. San Francisco: Friends of the Earth, 1981.

"Heavy Preference Tide Running Against Peripheral Canal," (rel. #1176). San Francisco: The Field Institute, June 4, 1982.

Helm, Michael, ed. *City Country Miners: Some Northern California Veins.* Berkeley: City Miner Books, 1981.

Jelinek, Lawrence J. *Harvest Empire: A History of California Agriculture.* San Francisco: Boyd and Fraser, 1979.

Kahrl, William, ed. *California Water Atlas.* Prepared by the Governor's Office of Planning and Research in cooperation with the California Department of Water Resources, Los Altos: William Kaufmann, Inc., 1978 and 1979.

Kahrl, William. *Water and Power: The Conflict over Los Angeles' Water Supply in the Owens Valley.* Berkeley and Los Angeles: University of California Press, 1982.

Kirsch, Jonathan. "Hot Water: Are We Really In It Without the Peripheral Canal?" *California,* May 1982.

Kirsch, Jonathan. "The Politics of Water." *New West,* September 10, 1979.

Magagnini, Stephen. "Why Peripheral Canal May Not Be Needed." *San Francisco Chronicle,* May 20, 1982.

Michels, Spencer. "California Split: Troubled Waters Over the Peripheral Canal Debate." *California Living,* April 11, 1982.

Nash, Roderick. *Wilderness and the American Mind* (rev. ed.). New Haven and London: Yale University Press, 1973.

Reisner, Marc. "Thirsty California: All This Water...Won't Be Enough." *GEO,* January 1981.

San Francisco Chronicle, April 26, May 11, May 20, June 1, June 3, June 4, June 9, June 10, and October 25, 1982.

Soble, Ronald L. "Foes of Canal Built Success on Two Points," *Los Angeles Times,* June 10, 1982.

Villarejo, Don. "Getting Bigger: Large Scale Farming in California." Davis, Calif.: California Institute for Rural Studies, 1980.

Villarejo, Don. "New Lands for Agriculture: The California State Water Project." Davis, Calif.: California Institute for Rural Studies, 1981.

Wiley, Peter and Robert Gottlieb. *Empires in the Sun: The Rise of the American West.* New York: G. P. Putnam's Sons, 1982.

Yonay, Ehud. "Paying the Piper: Notes on the Reclamation Reform Act of 1982." *California,* November 1982.

5. Imagine Two Californias

Berg, Peter and Raymond Dasmann. "Reinhabiting California." *Not Man Apart,* Mid-September, 1977.

Callenbach, Ernest. *Ecotopia.* Berkeley: Banyan Tree Books, 1975.

Callenbach, Ernest. *Ecotopia Emerging.* Berkeley: Banyan Tree Books, 1981.

Coyle, Wanda. *Fresno Bee.* May 2, 3, 5, and 6, 1982.

DeBell, Garrett, ed. *The New Environmental Handbook.* San Francisco: Friends of the Earth, 1980.

Department of Finance (State of California). "Population Estimates of California Cities and Counties. January 1, 1981, and January 1, 1982," May 1, 1982.

Eu, March Fong. "Report of Registration," (State of California), May 1982.

Eu, March Fong. "Statement of Vote," June 8, 1976; June 6, 1978; November 7, 1978; November 6, 1979; June 3, 1980; and June 8, 1982.

Field Institute, "The California Poll." Rel. #638, June 17, 1969.

Forbes, Jack. "Reconstituting California: A Model for Regional Self-Determination." *Raise the Stakes! The Planet Drum Review*, Winter 1981.

Kasindorf, Jeanie. "California Split." *New West*, May 9, 1977.

Keene, Barry. Address to the California Assembly, March 16, 1978.

Keene, Barry, Office of. Press release, March 16, 1978.

Lang, Joe. "Staff Analysis of Assembly Bill 2929." Government Organization Committee of the Assembly, March 1978.

Latham, Milton S. "Message from the Governor." *Journal of the Assembly, California 1860*.

Lawren, Bill. "California by the Slice." *Western's World*, November 1982.

Michels, Spencer. "California Split: Troubled Waters Over the Peripheral Canal Debate." *California Living*, April 11, 1982.

Michels, Spencer. "Two Californias." KQED's *Focus*, May 1981.

Reisner, Marc. "Thirsty California: All This Water. . .Won't Be Enough." *GEO*, January 1981.

Robinson, Frank M. and John Levin. *The Great Divide*. New York: Rawson, Wade Publishers, 1982.

San Francisco Chronicle, May 24, September 30, and October 1, 1982.

Sheridan, Peter. "Procedures for the Creation and Admission of States." The Congressional Research Service, The Library of Congress, September 24, 1980.

"Water Supply Contract Between the State of California Department of Water Resources and Kern County Water Agency." The Resources Agency of California, November 15, 1963.

Widney, J. P. "A Historical Sketch of the Movement for a Political Separation of the Two Californias, Northern and Southern, under both the Spanish and American Regimes." *Annual Publication of the Historical Society of Southern California 1888–9*, Vol. I, Los Angeles: Frank Cobler, "The Plain Printer," 1889.

Wiley, Peter and Robert Gottlieb. *Empires in the Sun: The Rise of the American West*. New York: G. P. Putnam's Sons, 1982.

Yeager, Kate. "Tomorrow: The Californias." 1978, sent to Barry Keene's office.

6. California Fusion—Healing the State

Arnold, Robert and Stephen Levy. *California Growth in the 1980s*. Palo Alto: Center for the Continuing Study of California's Economy, 1981.

Baker, George L. and Tom DeVries. "How the Big Oil Companies are Grabbing California's Water. *New West*, June 16, 1980.

Berg, Peter. *Reinhabiting a Separate Country*. San Francisco: Planet Drum Foundation, 1978.

"California's Water Future." Interim Hearing of the California Assembly Water, Parks, and Wildlife Committee, July 20, 1982.

Charter, S.P.R. "Will We Run Dry?" *Science Digest*, September 1982.

Delmatier, Royce, Clarence McIntosh, and Earl G. Waters, eds. *The Rumble of California Politics*. New York: John Wiley & Sons, 1970.

DeVries, Tom. "Never the Twain." *California*, November 1982.

Didion, Joan. *The White Album*. New York: Simon and Schuster, 1979.

Dvorin, Eugene P. and Arthur J. Misner, eds. *California Politics and Policies*, Reading, Calif.: Addison-Wesley, 1966.

"Effect of Conservation on Southern District Urban Water Demand for 1980, 1990, and 2000," Department of Water Resources, Southern District: State of California.

Garreau, Joel. *The Nine Nations of North America*. New York: Avon, 1981.

Kotkin, Joel, and Paul Grabowicz. *California Inc*. New York: Rawson Wade, 1982.

Kyink, Bernard L., Seyom Brown, and Ernest Thacker. *Politics and Government in California*. New York: Harper and Row, 1979.

Lantis, David. *California, Land of Contrast*. Belmont, Calif.: Wadsworth Publishing, 1970.

McWilliams, Carey. *The California Revolution*. New York: Grossman, 1968.

Phelps, Charles, et al. *Efficient Water Use in California:* Rand Corporation, 1978.

Powledge, Fred. "Water." *The Nation*, June 12, 1982.

Salzman, Ed. "Who Gets the Poppy?" *California Living*, July 18, 1982.

Stavins, Robert, and Phillip LeVeen. *A Guide to Proposition 9 for Concerned Citizens*. Berkeley: Public Interest Economics West, 1982.

Storper, Michael. "A Soft Water Path for California: Toward a Comprehensive Water Resources Policy." Contra Costa Water Agency, 1981.

Storper, Michael, and Richard Walker. "The California Water System: Another Round of Expansion?" Public Affairs Report, Bulletin of the Institute of Governmental Studies. Berkeley and Los Angeles: University of California.

Index

Other books by Island Press

Island Press is an independent nonprofit publisher with special interests in the environment and human experience. You may order any of the following titles directly from Island Press, Star Route 1, Box 38, Covelo, California 95428.

Tree Talk: The People and Politics of Timber, by Ray Raphael. Illustrations by Mark Livingston. $12.00

A probing analysis of modern forestry practices and philosophies. In a balanced and informed text, *Tree Talk* presents the views of loggers, environmentalists, timber industry executives, and forest farmers and goes beyond the politics of "production versus protection" to propose new ways to harvest trees and preserve forest habitats in a healthy economy and a thriving environment.

An Everyday History of Somewhere, by Ray Raphael. Illustrations by Mark Livingston. $8.00

This work of history and documentation embraces the life and work of ordinary people, from the Indians who inhabited the coastal forests of northern California to the loggers, tanbark strippers, and farmers who came after them. This loving look at history takes us in a full circle that leads to the everyday life of us all.

Pocket Flora of the Redwood Forest, by Dr. Rudolf W. Becking. Illustrations. $15.00

The most useful and comprehensive guidebook available for the plants of the world-famous redwood forest. Dr. Rudolf W. Becking, a noted botanist and Professor of Natural Resources, is also a gifted artist. The *Pocket Flora* includes detailed drawings, a complete key, and simple, accurate descriptions for 212 of the most common species of this unique plant community, as well as eight pages of color photographs. Plasticized cover for field use.

America Without Violence: Why Violence Persists and How You Can Stop It, by Michael N. Nagler. $8.00 paper. $13.00 cloth.

Challenging the widespread assumption that violence is an inevitable part of human existence, this book asserts that it *is* possible to live in a nonviolent society. The choice, Michael Nagler says, begins with each individual.

Wellspring: A Story from the Deep Country, by Barbara Dean. Illustrations. $6.00

The moving, first-person account of a contemporary woman's life at the edge of wilderness. Since 1971, Barbara Dean has lived in a handmade yurt on land she shares with fifteen friends.

The Trail North, by Hawk Greenway. Illustrations. $7.50

The summer adventure of a young man who traveled the spine of coastal mountains from California to Washington with only his horse for a companion. The book he has made from this journey reveals his coming of age as he studies, reflects, and greets the world that is awakening within and around him.

No Substitute for Madness: A Teacher, His Kids & The Lessons of Real Life, by Ron Jones. Illustrations. $8.00

Seven magnificent glimpses of life as it is. Ron Jones is a teacher with the gift of translating human beauty into words and knowing where to find it in the first place. This collection of true experiences includes "The Acorn People," the moving story of a summer camp for handicapped kids, and "The Third Wave," a harrowing experiment in Nazi training in a high school class—both of which were adapted for television movies.

Building an Ark: Tools for the Preservation of Natural Diversity Through Land Protection, by Phillip M. Hoose. Illustrations. $12.00

The author is national protection planner for the Nature Conservancy, and this book presents a comprehensive plan that can be used to identify and protect what remains of each state's natural ecological diversity.

Private Options: Tools and Concepts for Land Conservation, by Montana Land Reliance and Land Trust Exchange. $25.00

Practical methods for preserving land, based on the experience of thirty experts. Topics include income tax incentives for preserving agricultural land; marketing land conservation; management of conserved land areas; and the real estate business as a conservation tool.

The Conservation Easement in California, by Thomas S. Barrett and Putnam Livermore for The Trust for Public Land. $24.95 paper. $44.95 cloth.

An authoritative handbook for the preparation of easements in California, written by attorneys expert in conservation law. Includes discussion of the historical and legal background of easement techniques, state and federal tax implications, and solutions to the most difficult problems of drafting agreements.

The Christmas Coat, by Ron Jones. Illustrations. $4.00

A contemporary fable of a mysterious Christmas gift and a father's search for the sender, which takes him to his wife, his son, and his memories of big band and ballroom days.

Please enclose $1.00 with each order for postage and handling.
California residents, add 6% sales tax.
A catalog of current and forthcoming titles is available free of charge.